J. Lorenz (ed.)

3-Dimensional Process Simulation

Springer-Verlag Wien New York

J. Lorenz
Fraunhofer-Institut für Integrierte Schaltungen
Bauelementetechnologie
Schottkystrasse 10
D-91058 Erlangen

© 1995 Springer-Verlag/Wien
Softcover reprint of the hardcover 1st edition 1995

Typesetting: Camera ready by authors

Printed on acid-free and chlorine free bleached paper

With 164 Figures

ISBN-13:978-3-7091-7430-2   e-ISBN-13:978-3-7091-6905-6
DOI: 10.1007/978-3-7091-6905-6

EDITORIAL

This volume contains the Proceedings of the International "Workshop on 3D Process Simulation" which was held at the Campus of the University of Erlangen-Nuremberg in Erlangen on September 5, 1995, in conjunction with the 6th International Conference on "Simulation of Semiconductor Devices and Processes" (SISDEP '95). The workshop follows the International "Workshop on Technology CAD Systems" which was held at the Technical University of Vienna in conjunction with SISDEP '93.

Two-dimensional process simulation has achieved a certain degree of maturity. In consequence, well established program systems are available both via commercial software houses and from research institutes or universities. However, considerable improvements are still required in terms of the accuracy of the models used and the algorithms implemented, especially for adaptive meshing in case of time-dependent device geometries. In contrast to this, three-dimensional process simulation is a newly emerging field in which most efforts are dedicated to necessary basic developments. Technological demands from novel ULSI technologies require the availability of advanced three-dimensional process simulation tools to provide appropriate input to three-dimensional device simulation and, in this way, to allow for the support of ULSI device development and optimization. Major research and development activities dedicated to three-dimensional process simulation are presently being carried out around the world. In consequence, it can be expected that three-dimensional process simulation will prove to be a key component in advanced TCAD systems within the next few years.

Within the "Workshop on 3D Process Simulation", an overview of the activities being carried out in Japan, the USA, and Europe was given. The workshop consisted of ten invited presentations, among which four were from leading semiconductor companies, six were from research Centers of Excellence. Nine of these presentations are included in this book. In addition, a panel discussion on "Future Requirements and Trends in Multidimensional TCAD Simulation" was held during the Workshop.

The proceedings were printed from the authors' camera-ready manuscripts. I would like to express my sincere appreciation to the authors for their high quality contributions, their cooperation and efforts. Furthermore, I would like to thank my colleagues at FhG-IIS-B, especially F. Meyer, S. List, M. Schäfer, and C. Scordo, for their assistance in finalizing the proceeedings and organizing the Workshop.

Jürgen Lorenz
Editor

September 1995

SUPPORTING ORGANIZATIONS

Bayerische Verwaltung der Staatlichen Schlösser, Gärten und Seen
Bayerisches Staatsministerium für Wirtschaft, Verkehr und Technologie
VDE/VDI-Gesellschaft Mikroelektronik (GME)
IEEE Electron Devices Society
IEEE German Section
Informationstechnische Gesellschaft (ITG)
Siemens AG
Universität Erlangen-Nürnberg

SISDEP CONFERENCE COMMITTEE

G. Baccarani	Università di Bologna	ITALY
K. de Meyer	IMEC	BELGIUM
W. Fichtner	ETH Zürich	SWITZERLAND
M. Fukuma	NEC	JAPAN
H. Jacobs	Siemens	GERMANY
S. Laux	IBM	USA
C. Lombardi	SGS-Thompson	ITALY
P. Mole	BNR Europe	UNITED KINGDOM
M. Orlowski	Motorla	USA
A. Poncet	CNET/CNS	FRANCE
H. Ryssel	Universität Erlangen-Nürnberg	GERMANY
W. Schilders	Philips	THE NETHERLANDS
S. Selberherr	Technische Universität Wien	AUSTRIA
T. Toyabe	Hitachi	JAPAN
H. Van der Vorst	Rijksuniversiteit Utrecht	THE NETHERLANDS

SISDEP LOCAL ORGANIZING COMMITTEE

| M. Ebner | S. List | P. Pichler | M. Schäfer |
| T. Klauser | F. Meyer | H. Ryssel | C. Scordo |

Table of Contents

    M. Orlowski
        Advanced Products and Development Laboratory, Motorola Inc., Austin,
        Texas, USA

Three-Dimensional Topography Simulator: 3D-MULSS and Its Applications

Masato Fujinaga, Norihiko Kotani

ULSI Laboratory, Mitsubishi Electric Corporation,
1 Mizuhara 4-chome, Itami, Hyogo 664, JAPAN

Abstract

This paper introduces a three-dimensional topography simulator:3D-MULSS
and its applications. We focus on the description of the material surface and
the surface advancement, and then we present a 3D-MULSS model, with consideration to the probe size of observation, and based on the integration formula of the balance equation. Next, we show the simulation results of the
3D-MULSS: isotropic deposition, Aluminum-sputter deposition, isotropic etching, anisotropic etching, BPSG flow and sequential process steps. These results
make the accuracy of the 3D-MULSS clear, and also show that it is possible to
stably simulate the sequential process steps.

1. Introduction

The high integration of ULSI circuits has progressed, and devices of great complication
and very fine structure have been developed. In this situation, topography simulations are useful as a tool of LSI development to predict theoretically and quickly the
structure and the topography of devices.

At present, several topography simulators have been developed [1-5]. Particularly in
recent years, three-dimensional simulation has been investigated [6-12] and most of
these investigations have focused on mathematical algorithms in the propagation of
topography, which were based on the string model [14], the ray-tracing model[15],
and the cell model [16]. However, these papers do not discuss the physical aspects of
the material topography. By 'physical',we mean observation, which is distinguished
from 'mathematical'.

The LSI devices are fabricated by LSI processes such as deposition and etching,
which are controlled by mass-transport of atoms and molecules. From the microscopic
viewpoint, it is difficult to fix the location of the atoms and the molecules on the
material surface in the process of deposition or etching, because they are active,
vibrating and diffused. However, these phenomena depend upon temperature and
pressure. These phenomena are different from that of hard materials we know well.

Generally, the numerical calculation of mass-transport for gas and liquid is based on
the integration formula of the balance equation, of which the algorithm is that small
regions are specified, flux into or out of each region is calculated, and the material
quantity in the region is derived at short time intervals. Then, the material quantity

is described as an average value such as concentration, density or probability of existence. The string model doesn't have all of the above algorithm. The cell model dose have this algorithm, but cannot obtain oblique surfaces or smooth surfaces. In order to obtain a model which satisfies with both the algorithm and the smooth surface, the physical meaning of the macroscopic topography needs to be made clear. By making it clear, we have been able to develop a new algorithm and three-dimensional topography simulator:3D-MULSS (MUlti Layer Shape Simulator) [13].

In this paper, first, we discuss some merits and difficulties of the two typical conventional models: the string model and the cell removal model. Next, we present a physical description of the material topography, and a new algorithm of surface advancement and mass-transport for LSI fabrication processes such as etching and deposition. In section 4, we show the simulation results with this algorithm. In section 5, we discuss the difficulties of topography simulations. Finally, we present the several important points of topography simulations and summarize the 3D-MULSS and its applications.

2. The conventional model

2.1. The string model

In the string model [17], material boundary (topography) is approximated by a series of points which is joined by straight line segments, as shown in Fig. 1. These segments are moved in such a way that the local speed of each segment is the function of the local points and time. In the case of a 3D-model, the straight line segments are changed into triangles or polygons. In the string model, the algorithm of the propagation is easy to understand, and it is widely used in two-dimensions. But, there are some difficult problems concerning the control of the segments. Particularly in three-dimensions, it is very difficult to control the segment size and to delete the loops of segments [9], The string model is very useful for the analysis of stress and strain of hard materials such as a bent board, because the shape of the hard board is very little changed and the topology isn't changed at all. But it is difficult to apply the movement of the hard material surface to the propagation of topography which is greatly changed due to the mass-transport of atoms or molecules.

LSI fabrication processes such as etching and deposition are controlled by the mass-transport of atoms or molecules. When we calculate the mass-transport of gas or liquid numerically, we generally use the algorithm as follows:
1) a small region is specified,
2) the flux into and out of the regions is calculated,
3) the densities of the materials in the region are derived at short time intervals.
But in the case of the string model, only the segments or the points are only moved, and the region isn't specified and the density isn't calculated. As a result, we need to take care to advance the segments, in case there is an acute angle shape. Even if the segments are very small in the more accurate calculation, a new problem arises, which is related to the calculation of surface area and curvature. When we consider the novel case shown in Fig. 2, two shapes (a) and (b) seem to be the same if $N \rightarrow \infty$, but the surface areas are different mathematically. This means that the very small segments don't lead to accurate calculation of surface area.

The ray-tracing model also has the same problems as the string model.

2.2. The cell removal model

In the cell removal model [7], an analysis region is divided into cubic cells, and then the surface cells of the etched material are removed at a speed proportional to the surface area of the etched material, as shown in Fig. 3. In the cell model, it is easy to deal with mass-transport if the cell size is small enough. But it is difficult to describe arbitrary oblique surface and curvature.

Although it is well known that many cells are needed for high resolution, particularly in a 3D-model, previous papers have not discussed how to derive the oblique surface in the cell model. For example, when the surface area is calculated to derive the storage node's capacity of DRAM (Dynamic Random Access Memory), it is necessary to derive the oblique surface accurately. As shown in Fig. 4, the length of the oblique line is $\sqrt{2}$, but the length of the surface becomes 2 because it is expressed as only squares even if the number of cells increases. But if the number of cells is increased enough, an oblique surface of 45° angle can be seen with the naked eye. This leads to the observation described in Section 3. In addition, it is necessary to derive the curvature of the material surface for the topography simulation of BPSG (BoroPhospho Silicate Glass) flow. In this case, it is almost impossible to simulate the BPSG flow accurately with this cell model.

3. 3D-MULSS's Model

3.1. A physical description of material topography

In this section, let's think about the macroscopic topography. When $C(x, y, z)$ is assumed to be the distribution of material density as illustrated in Fig. 5, the density is constant in the inside and outside of the material, and it can be given as follows:

$$C(x, y, z) = \left\{ \begin{array}{ll} C_0 & \text{(inside)} \\ 0 & \text{(outside)} \end{array} \right. \qquad (1)$$

where x, y and z are assumed to be the axis of 3D-space. This equation(1) can be rewritten as

$$\nabla C(x, y, z) = 0. \qquad (2)$$

Therefore, the interface or surface can be defined as

$$\nabla C(x, y, z) \neq 0. \qquad (3)$$

This idea that the surface has thickness and the density is changed continuously is often used for the definition of the interface between a solid and a liquid (a gas). It has been said that the surface thickness is less than one or two layers (: about 10 Å) in the atomic levels under the condition that the temperature is lower than a certain critical point [18]. The material surface is the place where lights, electrons and ions begin to be scattered and the electrical potential is changed. From the atomic viewpoint, the local place where the lights are scattered is different from where the electrons and the atoms are scattered, because the mechanism of the scattering for the lights is different from those of electrons and atoms. The actual material topography is observed by the lights or the electrons, of which size is an important parameter.

In order to determine the normal direction of the material surface, particles and lights are hit against the material surface, and then the scattered particles are observed. We conclude that the normal vector of the material surface is the direction of the bisector between the incident angle and the recoil angle, as shown in Fig.6. Then, it must be assumed that the incident particles are larger than the atoms of the material, and the wave-length of the incident lights is as long as that of visible rays. This assumption means that the incident particles are large enough to neglect small concave and convex parts of the material surface, and corresponds to the assumption that the density can be differentiated.

As illustrated in Fig. 6, let's consider that the incident particles hit the material surface. Then, if the material density of the points (A) and (B) are equal, the magnitudes of the forces (f_A) and (f_B) are equal, and only components of those forces parallel to $-\nabla C$ remain. Therefore, in the case of elastic scattering, the direction of the bisector between the incident angle and the recoil angle is equal to that of $-\nabla C$. The force to the particles is an average force of the many atoms. Additionally, we can also regard C(x, y, z) as a physical value such as the electric potential, the index of light, the permitivity, and so on. For example, if C is assumed to be the electrical potential, $-\nabla C$ is equivalent to an electrical force. When the lights hit the surface, the scattering mechanism is related to the distribution of the electron density. The well known mirror law means that the lights are reflected to the same direction as the particles of the elastic scattering. As described above, the material surface means the place in which the density $C(x, y, z)$ meet the inequality (3). When a small surface region is considered, $-\nabla C$ is the normal vector on the small surface.

The thickness of the interface is the order of Å, but if the probe of the observation is larger than the thickness, we can't measure the thickness more accurately than the probe size. The optical lithography demonstrates this fact [19] [20]. The image of the mask pattern is given by the optical intensity, of which the contour line (the equi-intensity line) becomes the image of the mask pattern. The macroscopic topography depends on the probe size of the observation, whatever the procedure of the observation. If the probe size is not determined, we can't obtain the macroscopic topography.

3.2. A new algorithm of the surface advancement

Of the various LSI processes, it is etching and deposition that fabricate hard devices; however, the device surface is active during the fabrication. Even if the position of each atom or molecule is indeterminate, the mass is conserved. We introduce a new simulation method of mass-transport in etching and deposition, which is a new algorithm of surface advancement. Let us call small regions "cells". Then, we assume that the positions of atoms and molecules are indeterminant in each cell and the size of the cell is equal to the probe size of the observation. This assumption is important and is discussed in section 5. The mass-conservation can be given by the integrated formula of the balance equation:

$$\frac{d}{dt} \int \int \int_V C dV = - \int \int_S \boldsymbol{J} \cdot \boldsymbol{n} dS \qquad (4)$$

where C is material density, t is time, J is flux density, n is the normal vector of a small surface(dS).

In the cases of deposition and etching, a way to solve Eq.(4), is to
1) divide the area under investigation into small cells (small rectangular prisms) which have the normalized density (volume rate) of the material at the centers as illustrated in Fig 7,
2) give the initial distribution of C,

3) derive the material surface as the place of the equi-volume rate 0.5 in each cell by linear interpolation as shown in Appendix (A),
4) calculate the flux into or out of the surface of each small cell, and then
5) derive the material volume rate of each cell at short time intervals,
as shown below.

$$C_m^{(n+1)}(i,j,k)V(i,j,k)C_{m0} = C_m^{(n)}(i,j,k)V(i,j,k)C_{m0} + R_m^{(n)}(i,j,k)\triangle t^{(n)} \quad (5)$$

$$C_t^{(n)}(i,j,k) = \sum_m C_m^{(n)}(i,j,k) \quad (6)$$

$$C_m^{(n)} = \frac{C^{(n)}(i,j,k)}{C_{m0}} \quad (7)$$

Where:

C_m	:	volume rate (normalized density of the material(m))
(i,j,k)	:	cell numbers for x,y,z-directions
C_{m0}	:	density of the material(m)
$V(i,j,k)$	:	volume of cell(i,j,k)
C_t	:	total normalized density
$R_m(i,j,k)$	:	flux rate of the material into or out of the surface of the cell(i,j,k)
$C(i,j,k)$	:	density of the material(m) = the number of the particles(m)/$V(i,j,k)$ in the cell(i,j,k)
m	:	material index (Si=1, Polysilicon=2, ... resist=10,.....)
n	:	time step number
i,j,k	:	cell numbers for x,y,z-directions.

In the cases of etching and deposition, basically, the short time interval is defined as the period which is needed to change 1/2 of the cell's volume rate in one step.
6) After Eq.(6-7) are calculated, C_m and C_t are controlled at each time step in the cases of isotropic deposition and isotropic etching as follows:
Control (a) : If $C_t^{(n+1)}(i,j,k) > 1$, the surplus ($C_t^{(n+1)}(i,j,k) - 1$) is divided into its neighboring cells for which $C_t < 1$, proportionally to the contact area of the cell(i,j,k) and its neighboring cells. The surplus is deducted from $C_m(i,j,k)$.
Control (b) : If $C_m(i,j,k) < 0$, the deficit $C_m(i,j,k)$ is divided into its neighboring cells for which $C_m > 0$, proportionally to the contact area of the cell(i,j,k) and its neighboring cells. $C_m(i,j,k)$ is assumed to be zero.
In anisotropic cases, it is effective to divide the surplus and the deficit into the cells of the anisotropic direction.

In this case, the cell size is equivalent to the probe size of the observation. We use linear interpolation for the derivation of the equi-volume rate because it is very simple.
In the above description, the surface was defined as the equi-volume rate of 0.5. This is very suitable to the actual simulation. If the surface isn't defined as 0.5 (for example, the surface is defined as 0.3 or 0.8), a problem is generated in which there are duplicate or interstice regions between two different materials, as illustrated in Fig. 8. In general physics, the interface between the two different phases such as gas and liquid is defined as the place which is neither the gas phase or the liquid phase [], and the value of 0.5 can be derived from the definition of the interface as shown in Appendix(B). Therefore, the value of 0.5, being exactly half way, is most suitable.

In the new model, the place of the surface can be moved by changing the volume rate continuously, but there is a difference between the point of the equi-volume rate(0.5) and the ideal surface point. As shown in Fig. 9, if it is assumed that the widths of cell(1), cell(2) and cell(3) are d_1, d_2 and d_3, respectively, and the volume rates of cell(1), cell(2) and cell(3) are 1, x, and 0, respectively, the point(P') of the equi-volume rate:a can be given by

$$P' = \begin{cases} \left(\frac{1-a}{1-x}\right)\left(\frac{d_1+d_2}{2}\right) & (x \leq a) \\ \left(\frac{d_1+d_2}{2}\right) + \left(\frac{x-a}{x-0}\right)\left(\frac{d_2+d_3}{2}\right) & (a < x \leq 1) \end{cases} \qquad (8)$$

while the ideal surface point (P) can be given by

$$P = \frac{d_1}{2} + \frac{x}{d_2} \qquad (9)$$

Therefore, the difference between these points can be written as the function of the concentration (x) as follows:

$$F(x) = |P' - P|. \qquad (10)$$

The examples of the function $F(x)$ are shown in Fig. 10 (a), (b) and (c), which correspond to a=0.2, 0.5 and 0.8, respectively. In Fig. 10, the cell widths d_1, d_2, d_3 are assumed to satisfy the following equations:

$$\frac{d_1}{d_2} = \frac{d_2}{d_3} \qquad (11)$$

$$d_2 = 1. \qquad (12)$$

The dotted, the dashed and the solid lines correspond to the cases of d_1 =4.0, 1.0 and 0.2, respectively. As shown in these graphs, if $a = 0.2$, the function $F(x)$ of $d_1 = 4.0$ becomes smallest. If $a = 0.5$, $F(x)$ of $d_1 = 1$ becomes smallest. If $a = 0.8$, $F(x)$ of $d_1 = 0.2$ becomes smallest. In the case of the uniform mesh ($d_1 = 1$) and $a = 0.5$, the average of $F(x)$ is smaller than any other cases. In the case of the nonuniform mesh, $a = 0.2$ or 0.8 is suitable. But when the analysis region is divided, it is impossible to use the division of $d_1 = 0.2$ or 0.8 for all of the analysis region. The division of the analysis region is based on the uniform mesh, which is most often used. Therefore, $a = 0.5$ is most suitable from the viewpoint of accuracy. In the case of the uniform mesh and $a = 0.5$, the error F(x) has the maximum value :

$$F(x) = 0.089 \sim 1/10 \qquad (13)$$

$$x = 1 - \frac{\sqrt{2}}{2} \sim 0.293. \qquad (14)$$

Therefore, it is concluded that the accuracy of the new model is about 1/10 of the cell size, if the uniform mesh is used.

4. Applications

4.1. Isotropic deposition

The SEM photograph of TEOS by LPCVD (low pressure CVD) is shown in Fig. 11. It is a isotropic deposition and the circular coverage on the step can be seen. In the simulation of the isotropic deposition, the flux into a cell is proportional to the surface area in the cell. The surface is described as the equi-volume rate of 0.5. A 3D-simulation result of isotropic deposition is shown in Fig.12. Fig.12(a) is a 3D-view, and Fig.12(b) and Fig.12(c) are an A-A' cross section. Fig.12(b) is the simulation result using the method(1), and Fig.12(c) is that using the method(2), where the method(1) and (2) are methods to derive the contour surface of an equi-volume rate, and are described in Appendix(A). The calculation times of Fig12(b) and Fig.12(c) are 883 seconds and 149 sec using HP9000(735), respectively. The number of grids are $100 \times 50 \times 40$ and the uniform grids are used. The calculation with method(2) is faster than that of method(1), and the accuracy of method(2) is almost same as that of method(1). Therefore, method(2) is practical, and so it is used for simulations from here on.

The distribution of TEOS volume rates is shown in Fig. 13. Particularly, we can see that the volume rates at the lower corner of the step is more than 0.8 and less than 1 in the material, regardless of the inside of the material. This result is related to the fact that the accuracy is 1/10 of cell size. When the TEOS is etched in this state, the TEOS at the corner is fast etched. If the volume rate are problem, the following may be added:

if a cell doesn't have the surface of the equi-volume rate 0.5 and its volume rate is more than 0.5 and less than 1.0, its volume rate can be assumed to be 1.

Thus, it is easy to make these volume rates unity, if the small error of mass-conservation can be ignored.

4.2. Sputter deposition of aluminum

The position of the wafer with respect to the circular Al-target which has erosion areas is illustrated in Fig. 14. There is a plasma of argon between the wafer and the target. Al-atoms are sputtered by argon-ions hitting the Al-target. The erosion area of the Al-target is illustrated in Fig.14 with the bold lines. The angular distribution of the sputtered aluminum is given by $cos(\theta)$. Because of the low gas pressure, it is assumed that the mean free path of Al-species is long enough to hit the wafer surface. The sticking coefficient of Al to the wafer surface is assumed to be 1. The deposited aluminum atoms are always diffused at least in one cell's width, as long as our model is used. Then, it is assumed that the density of the deposited film is constant.

A 3D-simulation result is shown in Fig. 15. The number of grids are are $50 \times 50 \times 44$, and the calculation time are 3084 sec using CRAY-Y/MP. The coverage of Aluminum film on the bottom and on the side-wall of the via hole is not good, because of the surface shadowing. In the case of 3D-simulation for arbitrary topography, it takes a long time to calculate the surface shadowing. As illustrated in Fig.16, the calculation is given by considering the direction of (θ, ϕ) from a point on the surface, and determining whether Al-species come from that direction to the surface.

As θ and ϕ are satisfied with $0 < \pi/2$ and $0 < 2\pi$, respectively, they are divided into N and $4N$, respectively. If the straight line of the direction (θ, ϕ) passes through the cells of materials, Al-species doesn't come from the direction (θ, ϕ). Appendix(C) describes an algorithm to find the cells through which a straight line passes.

4.3. Isotropic etching

In the simulation of isotropic etching, it is assumed that the flux of etched material which flow out of a cell is proportional to the surface area in the cell which has the equi-volume rate of 0.5. As an example of isotropic etching, a SEM photograph of the

dry etching of Si_3N_4 film is shown in Fig. 17. This etching is strongly controlled by
the chemical reaction on the surface. A 3D-simulation of isotropic etching is shown
in Fig.18. The number of grids are $75 \times 75 \times 80$, and the calculation time is 712 sec
using HP9000(735). Particularly, it is not good to leave a little material in the air,
even if the accuracy is 1/10 of cell width. Therefore, if a cell has a volume rate of
less than 0.5 and there is no cell that has a volume rate of more than 0.5 around the
cell, the volume rate of the cell should be made 0.

4.4. Anisotropic etching

In the simulation of anisotropic etching, it is assumed that the flux of etched material
which flows out of a cell is proportional to the flux of an etchant which enters the
surface in a cell. The surface shadowing effect was considered because the etchant
moves in a straight line. For an example of experimental data, a SEM photograph of
anisotropic etching is shown in Fig.19. The 3D-simulation result is shown in Fig.20.
The number of grids is $75 \times 75 \times 80$, and the calculation time is 823 sec, using
HP9000(735). Here, the important point is whether the algorithm of surface advancement which is similar to the isotropic case can be applied to anisotropic etching. It
is found that it is possible by comparing the experiment and the simulation. In the
case of the small enough cells, the control(B) which is equivalent to the isotropic case
can be used. But in the case of the larger cells, the problem occurs that the edge
topography is changed. In order to solve the problem in the case of larger cells, the
control(B) may be changed as follows:
Control (B'): If $C_m(i, j, k) < 0$, the deficit $C_m(i, j, k)$ is divided into the neighboring
cell under it.
As a result, we can see the anisotropic topography as a matter of convenience even if
the cell size is large.

4.5. BPSG flow

The viscous flow model and the surface diffusion model have been presented as BPSG
flow model [21] [22]. In our 3D-MULSS we use the surface diffusion model, however,
which is a phenomenological model. As the BPSG flow is caused by surface tension,
whichever model is used, 3D-curvature must be derived at the local area of the surface.
In our 3D-MULSS [23], the 3D-curvature is given by the $\frac{dA}{dV}$, where dA and dV are the
increment of the surface area and the volume, respectively. dA and dV are calculated
as follows:

As illustrated in Fig.21, if we look at triangles ABC(i) and DBC(j), we see that if
those triangles are shifted to normal directions for a distance ϵ, the previous triangles
are separated and an interstice or a duplication is generated between the two triangles.
The interstice or duplication is the increment dA_{ij} of the surface area. In the case of
no interstice or duplication, the reciprocal of the curvature radius is 0. In the case
of interstice, the area increment is positive. In the case of duplication, it is negative.
Corresponding to the area increment, the volume increment dV_{ij} is given as the area
of triangles G_iBC and G_jBC multiplied by the distance ϵ, where the points G_i is the
mass center (or inside center) of the triangle (i) and the point G_j is that of the triangle
(j).

Therefore, the curvature of a surface cell can be written as shown below.

$$dA = \sum_{i<j} \vec{b}_{ij} \times (\vec{n}_i - \vec{n}_j)\epsilon \qquad (15)$$

$$dV = \sum_{i<j} U_{ij} \epsilon \tag{16}$$

$$dA/dV = \frac{\sum_{i<j} \vec{b}_{ij} \times (\vec{n}_i - \vec{n}_j)}{\sum_{i<j} U_{ij}} \tag{17}$$

where it is assumed that $\vec{n}_i$ and $\vec{n}_j$ are normal unit vectors of triangles (i and j), $\vec{b}_{ij}$ is a vector of side(BC), $\times$ is outer product, U_{ij} is the area of the triangles $(G_i BC + G_j BC)$, $\sum_{i<j}$ is the summation of combinations of triangles in a cell. Both $\vec{n}_i$ and $\vec{n}_j$ are unit vectors, and so b_{ij} is always perpendicular to $(\vec{n}_i - \vec{n}_j)$. Thus, the summation of formation (3) and (5) considers the magnitude of outer product and only the sign for the direction of $\vec{n}_i$.

The 3D-simulation result of BPSG flow is shown in Fig.22, of which the calculation time is 995 sec with CRAY-Y/MP.

4.6. Sequential process steps

LSI devices are fabricated by sequential process steps. In this paper, two simulations of sequential process steps are shown.

One is the topography simulation of comparatively small region($4\mu m \times 4\mu m \times 5\mu m$). which is shown in Fig. 23. The number of process steps is 26, and the calculation time is 384 sec using CRAY-Y/MP. Our 3D-MULSS can simulate sequential process steps stably.

The other is the simulation of concave and convex on the surface in a comparatively larger region, which is shown in Fig. 24. In the photolithography process, the developed resist shape depends on the concave and the convex of substratum, because of the absorption or reflection of light or defocus effects. We need to find the dangerous region on the mask data. Fig.24 shows the contour of the surface height after the formation of the interlayer dielectrics film (BPSG), in which there is a local concave at the position of $(x = 12\mu m, y = 7\mu m)$. The resist on the place of this concave becomes thick, and if a contact hole is formed at this place, the contact diameter becomes small. It leads to failure of devices. This calculation analysis region is $13\mu m \times 15\mu m \times 5\mu m$, the number of process steps is 30, the number of grids are $30 \times 26 \times 50$ and the calculation time is 427 sec with CRAY-Y/MP. If the calculation is speeded up 10 times, it will be possible to correct the mask patterns in larger regions such as 1mm $\times$ 1mm, based on the simulation result.

5. Discussion

Topography simulation has a difficulty in the calculation of the mass-transport in the moving-interface between a solid and a gas or a liquid, because an atom or a molecule in the bulk of a solid are fixed at a local region, but that in a gas or a liquid is not fixed. Generally, the analysis of the mass-transport in a gas or a liquid is that a small region is specified, the flux in and out of the regions is calculated, and the densities of the materials in the region are derived, based on the mass-conservation. We have thoroughly applied this algorithm to topography simulation. In this paper, a rectangular prism is used as the small region, but a rectangular prism isn't the only possibility. For instance, a tetrahedron can be used, too. The difference between the 3D-MULSS model and the cell model is not only that the volume rate of a cell is a continuous value from 0 to 1, but also that the volume rate is derived by calculating

the flux in or out of the material surface which is given by the equi-volume rate area(
0.5). This calculation method of flux is common to the string model and the 3D-
MULSS model. But the string model doesn't specify the small region. Therefore, it
can be said that the 3D-MULSS model is intermediate between the string model and
the cell model. In the case of 3D-MULSS model, when the interfaces of more than
three materials touch each other simultaneously, an interstice is generated at that
place. This interstice becomes smaller if the cell size is small.

The 3D-MULSS model makes the assumptions that the material atoms are diffused
and the growing direction is isotropic in a cell. In the cases of sputter-deposition
in low temperature and directional growth such as epitaxial growth related to the
crystal, the cell size must be considered. In this case, the morphological method
proposed by Stresser et al. [12] can be used. Then, the growth direction should be
given by microscopic status. The string model and the cell model also require this
care, as long as they are macroscopic models.

In addition, generally, microscopic surface topography is indeterminate. In this paper, we have not attempted descriptions of the micro scopic surface such as Fractal
structure.

6. Conclusions

First, the merits and the weak points of the string model and the cell model have been
discussed. The important point is to establish a technique both for obtaining the surface topography with arbitrary oblique surface and for calculating the mass-transport
mechanism accurately in the moving boundary. The string model can deal with the
oblique surface, but it is difficult to calculate the mass-transport accurately because
the small region to specify is not clear. Additionally, when the 2D-string model is
expanded to a 3D-model, complicated controls of the facets are needed in the algorithm. The cell model can calculate the mass-transport accurately, but it is difficult
to obtain smooth surfaces such as the oblique surfaces and the curvature. In order to
satisfy these two conditions simultaneously, we need to distinguish the macroscopic
topography from the microscopic topography and to determine the macro-scopic topography consistently by giving the procedure of the observation. The macroscopic
topography of materials depends on the state of the surface atoms and the probe size
of the observation. By considering the above, we have proposed a simple algorithm
of topography simulation, based on the integration formula of the balance equation.
It is described as follows:

1. divide the area under investigation into small cells which have the normalized
   density(volume rate) of the material at the centers of cells,

2. give the initial distribution of volume rate,

3. derive the material surface as the area of the equi-volume rate 0.5 in each cell
   by the linear interpolation,

4. calculate flux into or out of the surface of each small cell, and then

5. derive the material volume rate of each cell at short time intervals.

This algorithm is implemented in our 3D-MULSS. In particular, we have to take care
to calculate flux into or out of the material surface, which is described as the equi-
volume rate : 0.5. The accuracy of the simulation is about 1/10 of a cell's width.

Then, it is shown by the definition of the material interface and the consideration of the error that the equi-volume rate 0.5 is most suitable to this topography simulation.

Our 3D-MULSS was applied to isotropic deposition, aluminum sputter deposition, isotropic etching, anisotropic etching, BPSG flow and finally the sequential processes. We compared the simulation results and the experimental results for the each individual process, and found the good agreement between the simulations and the experiments. In particular, this algorithm is found to be effective for the isotropic growth. We also found it possible to the very stable calculations of sequential processes in three-dimensions. However, when this algorithm is applied to processes such as the directional growth related to microscopic crystal, we should give care to the cell size.

Acknowledgments

We wish to thank S. Sakamori, A. Fujii, A. Ohsaki and H. Honda of ULSI Laboratory, Mitsubishi Electric Corporation for providing experimental data.

appendix

7. Calculation of 3D-contour surface

A1. Accurate calculation

As illustrated in Fig.25, an analysis region is divided into cubic cells. The volume rate of material is given in the center of each cubic cell. First, the volume rates at the vertexes are derived from a linear interpolation of the eight values at the cell centers as follows:

$$C(x,y,z) = \sum_{i=1}^{8} \Delta_i C_i \tag{18}$$

where it is assumed that

$$\Delta_i = \frac{V_i}{V_0} \tag{19}$$

V_0 is volume of a cubic cell, V_i is volume of the cubic which is specified by the point (x,y,z) and the opposite vertex of the cell center to the vertex (i), and C_i is the volume rate in the center of a cubic cell which is in the neighborhood of vertex (x,y,z). The cubic cell is then divided into 24 tetrahedrons, as illustrated in Fig.26. Next, the volume rate is derived similarly at the vertices of each tetrahedron, and if C_0 is assumed to be the interface volume rate(0.5), the points of C_0 on the sides of each tetrahedron are calculated by linear interpolation. If these points are connected with each other by straight lines in a tetrahedron, a triangle or polygon is derived. The surface of the equi-volume rate is constructed of these triangles and polygons.

A2. Fast calculation

The volume rates at the vertexes of a cubic cell are calculated in the same way as (A1). Then, the points of the volume rate C_0 on the sides of a cubic cell are derived by the linear interpolation. As illustrated in Fig.27, if these points are connected with each other by straight lines in the cubic cell, a triangle or a polygon is derived, which is surface of the equi-volume rate C_0.

8. Derivation of interface(0.5)

The theory of renormalization is applied to this problem. A state is extended by the transformation group of renormalization. An interface is defined as a place which is not solid (liquid) or gas. If the state is not changed by the transformation, it is an interface. As shown in Fig. 28, the region (A) is constructed of four small regions. It is assumed that no small region can have both solid and gas. The necessary condition that the extended region(A) becomes solid, are
(1) all four regions are solid,
or (2) three regions are solid and one region is gas,
or (3) if two regions are solid and two regions are gas, half is contributed to the probability.

As described above, three cases need to be considered. Here, it is assumed that the probability of solid for a small region is pi, that of gas is $q(= 1-p)$. The probability

$f(p)$ of solid for the extended region A can be written as follows:

$$f(p) = {}_4C_4p^4(1-p)^0 + {}_4C_3p^3(1-p)^1 + {}_4C_2p^2(1-p)^2(\frac{1}{2}) \qquad (20)$$

$$= -2p^3 + 3p^2 \qquad (21)$$

where nC_k is assumed to be a combination.

This function $f(p), (0 < p < 1)$ is illustrated in Fig. 29. Here, $f(p)$ is put into p and this calculation (21) is performed. This transformation is an infinite number of times performed.

- If $p < 0.5$, f approaches to zero:(gas).

- If $p > 0.5$, f approaches to one:(solid).

- If $p = 0.5$, f is 0.5, which is not changed no matter how many times the transformation is carried out.

That region of $p = 0.5$ is not solid or gas, and it is an interface. A simpler derivation is described as follows:

$$p = f(p) = -2p^3 + 3p^2 \qquad (22)$$
$$p(p-1)(2p-1) = 0 \qquad (23)$$
$$p = \frac{1}{2}, (0 < p < 1). \qquad (24)$$

9. The passing cells of a straight line vector

One of the four directions :$(x > 0, y > 0), (x < 0, y > 0), (x > 0, y < 0), (x < 0, y < 0)$ is determined for the vector of a straight line. It is assumed that the velocity vector $\vec{v}$ starts at the point A(i, j, k) in Fig. 30. If the vector $\vec{v}$ is the direction of $(x > 0, y > 0)$, the three pass times :t_x, t_y, t_z at the cell boundaries are calculated by increasing the cell numbers step by step. And then, it is determined which pass times are long or short. For example, if $t_x < t_y < t_z$, z-cell number k is made constant till t_z, y and z-cell numbers: j and k are made constant till t_y, and the passing cells are derived as increasing x-cell number:i. This problem is equivalent to the sorting algorithm of the times at which a vector passes through the cell boundaries. However, in the case of surface shadowing, it is better that if a material cell is met, the increment of the cell numbers (i, j, k) is stopped and then the passing cells are calculated for the next direction (θ, ϕ).

References

[1] 'Wafer topography simulation', Process and Device Simulation for MOS-VLSI Circuits, edited by P.Antognetti et al. Martinus Nijhoff Publishers, Boston, 411 1983.

[2] G. Oldham, S.N. Nandgaonkar,A.R. Neureuther, and M.M.O'Toole, "A general simulator for VLSI Lithography and etching processes : Part I - Application to Projectin lithography,"IEEE Trans. Electron Devices, Vol. ED-26, pp.717-722, 1979.

[3] G. Oldham, A.R. Neureuther, C.Sung, J.L.Reynolds and S.N. Nandgaonkar,IEEE Trans. Electron Devices, ED-27, p.1445, 1980.

[4] J.Pelka, et al. "Simulation of Dry Etch Process by COMPOSITE," IEEE Trans. CAD, CAD-7, 154 , 1988.

[5] J.Lorenz, et al. "COMPOSITE-A Complete Modeling Program of Silicon Technology," IEEE Trans. CAD, CAD-4, 421, 1985.

[6] A.Moniwa, et al.:"Three-Dimensional Photoresist Imaging Process Simulator for Strong Standing-Wave Effect", IEEE Trans. CAD, CAD-6 431 (1987).

[7] Y. Hirai et al : Symposium on VLSI Technology IEEE CAT. No.87th 0189-1, p.15, 1987.

[8] M. Fujinaga, N. Kotani, T. Kunikiyo, H. Oda, M. Shirahata, and Y. Akasaka, "Three-Dimensional Topography Simulation Model: Etching and Lithography," IEEE Trans. ED Vol-37, No. 10, 1990.

[9] E.W.Scheckler and A.R.Neureuther, "Models and Algorithms for Three-Dimensional Topography Simulation with SAMPLE-3D," IEEE Trans. CAD, Vol.13, No.2, Feb., p.219, 1994.

[10] S.Tazawa, F.A.Leon, G.D.Anderson, T.Abe, K.Saito, A.Yoshii, and D.L.Scharfetter, IEDM92 Technical Digest, IEEE, p.173, 1992.

[11] F.A.Leon, S.Tazawa, K.Saito, A.Yoshii, and D.L.Scharfetter, VPAD93 Digest, JSAP and IEEE, p.58, 1993.

[12] E.Stresser, K.Wimmer, and S.Selberherr, VPAD93 Digest, JSAP and IEEE, p.54, 1993.

[13] M. Fujinaga, et al., "New Topography Expression Model and 3D-Topography Simulation of Al-Sputter Deposition, Etching and Photolithography, " *IEDM Tech. Dig.,* pp. 905–908, 1990.

[14] R.E. Jewett, P.I. Hagouel, A.R. Neureuther, and T. Van Duzer, "Line-Profile Resisi Development Simulation Techniques " Polymer Engineering and Science, June, Vol. 17, No.6, p.381, 1977.

[15] P.I. Hagouel and A.R. Neureuther, ACS Organic Coatings and Plastics Chemistry 170th. meeting, Chicago, p.298, 1975.

[16] F.H. Dill, A.R.Neureuther, J.A. Tuttle, and E.J. Walker, "Modeling Projection Printing of Positive Photoresists " em IEEE Trans. Electron Devices, Vol. ED 22, pp. 456–464, 1975.

[17] W.Fichtner "Process Simulation," VLSI Technology, Edited by S.M.Sze, 1983, McGraw-Hill Book Company.

[18] J. W. Cahn and J. E. Hilliard, "Free Energy of a Nonuniform System. I. Interfacial Free Energy," J. Chemical Physics, Vol. 28, No. 2, pp. 258-267, 1958.

[19] B.J. Lin, " Partial Coherent Imaging in Two Dimensions and the Theretical Limits of Projection Printing in Microfabrication," IEEE Trans. Electron Devices, Vol. ED-27, No. 5, pp. 931-938, 1980.

[20] M. Yeung, " Modeling Aerial Images in Two and Three Dimensions,' Proceedings of the Kodak Microelectronics Seminar INTERFACE'85, pp.115-126, 1986.

[21] Francisco A. Leon, " Numerical Modeling of Glass Flow and Spin-on Planarizat ion," IEEE Trans. Electronon CAD, Vol. 7, No. 2,1988, pp. 168-173.

[22] H.Umimoto, S. Odanaka, S. Imai, "A Three-Dimensional Dynamic Simulation of B orophosphosilicate Glass Flow", Tech. Dig. of Symp. VLSI Tech., 10-4, pp.99-100, 1991.

[23] M. Fujinaga, I.Tottori, T.Kunikiyo, T.Uchida, N.Kotani, and K.Tsukamoto, "3D-numerical Modeling of Thermal Flow for Insulating Thin Film Using Surface Diffusion," pp. 631, Vol.14, No.5, 1995.

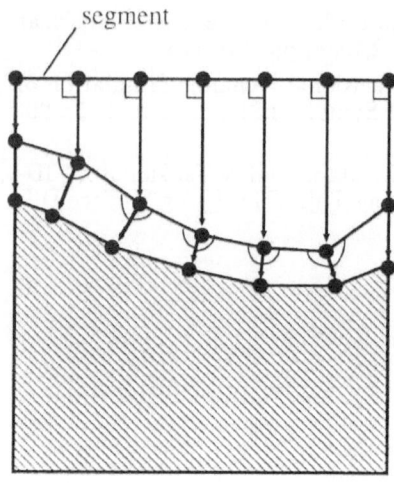

Fig.1 string model

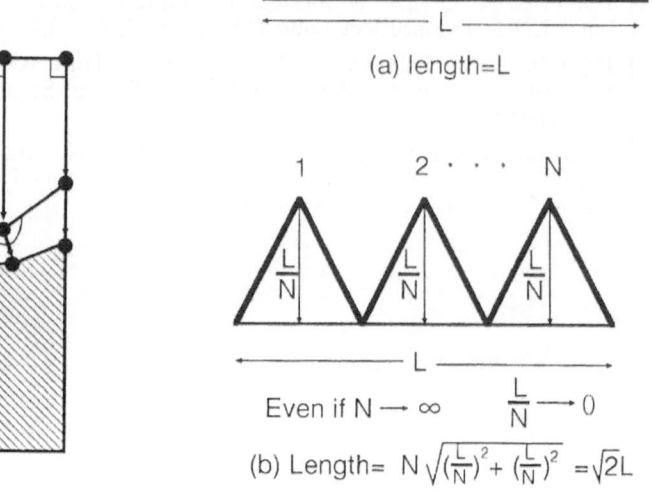

Fig.2 Different surface areas

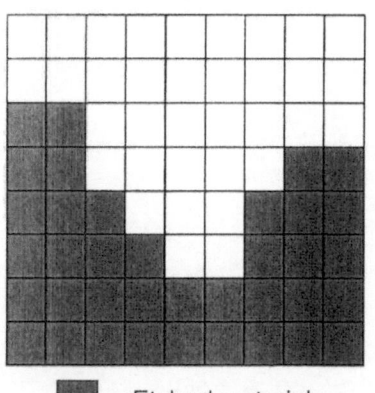

■ :Etched material

□ :Etchant (Liquid)

Fig.3 Cell removal model

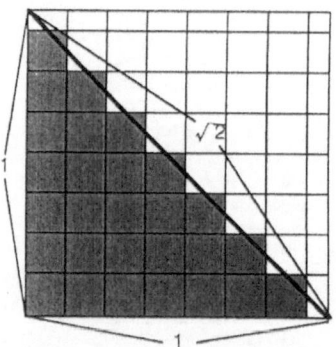

Fig.4 Surface area in cell model

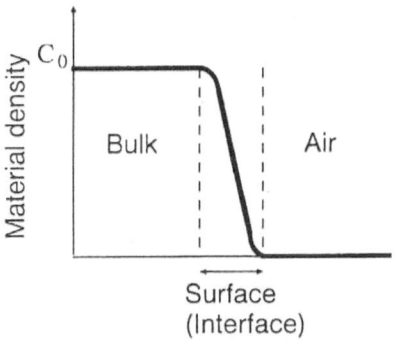

Fig.5 inside and surface

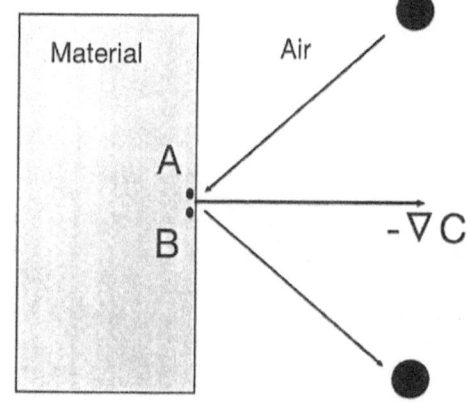

Fig.6 Observation of Surface

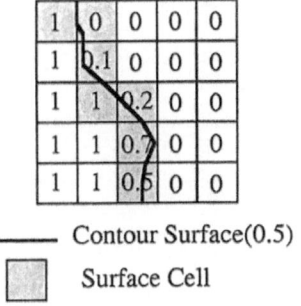

—— Contour Surface(0.5)

▨ Surface Cell

Fig. 7　Cell and Volume rate

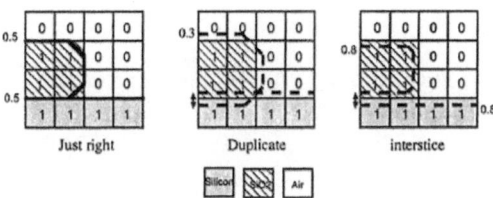

Fig.8 Equi-volume rate and interface

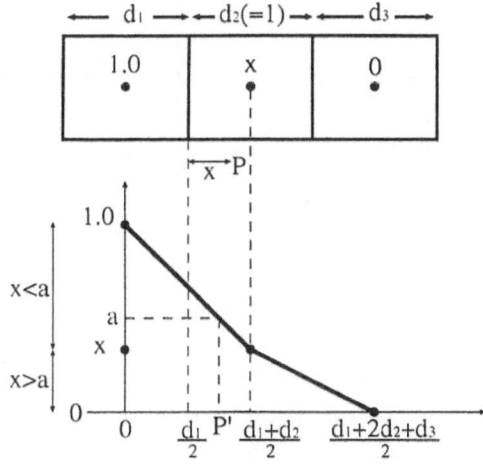

Fig.9 Difference of P and P'

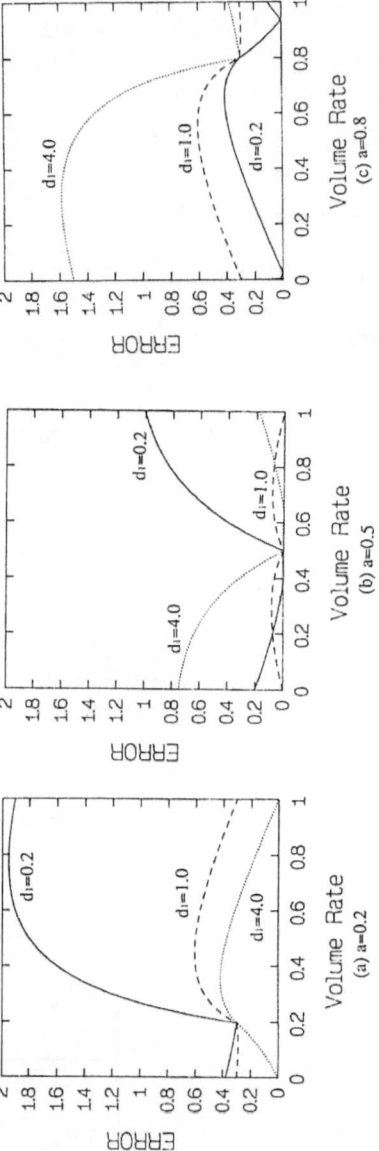

Fig.10 Examples of function F(x)

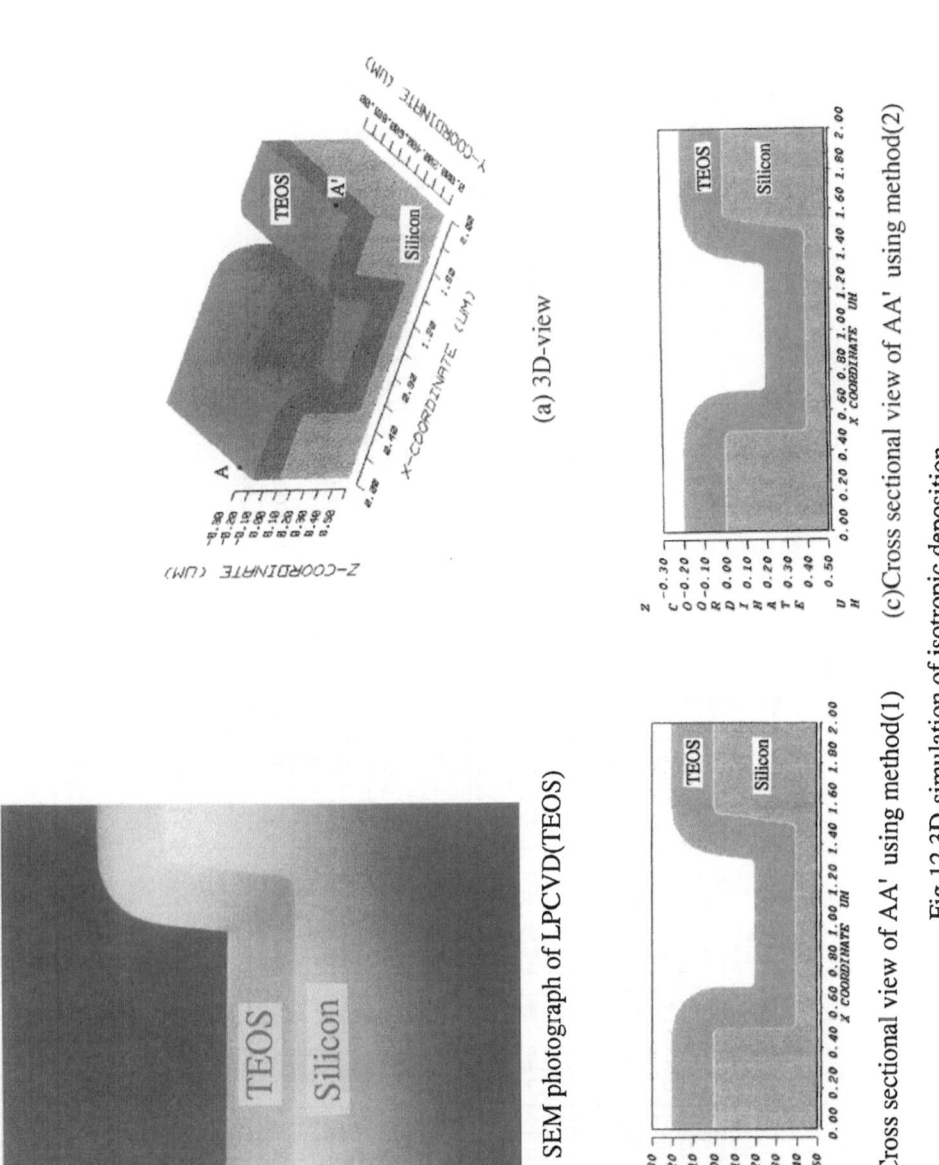

Fig.11 SEM photograph of LPCVD(TEOS)

(a) 3D-view

(b)Cross sectional view of AA' using method(1)

(c)Cross sectional view of AA' using method(2)

Fig.12 3D-simulation of isotropic deposition

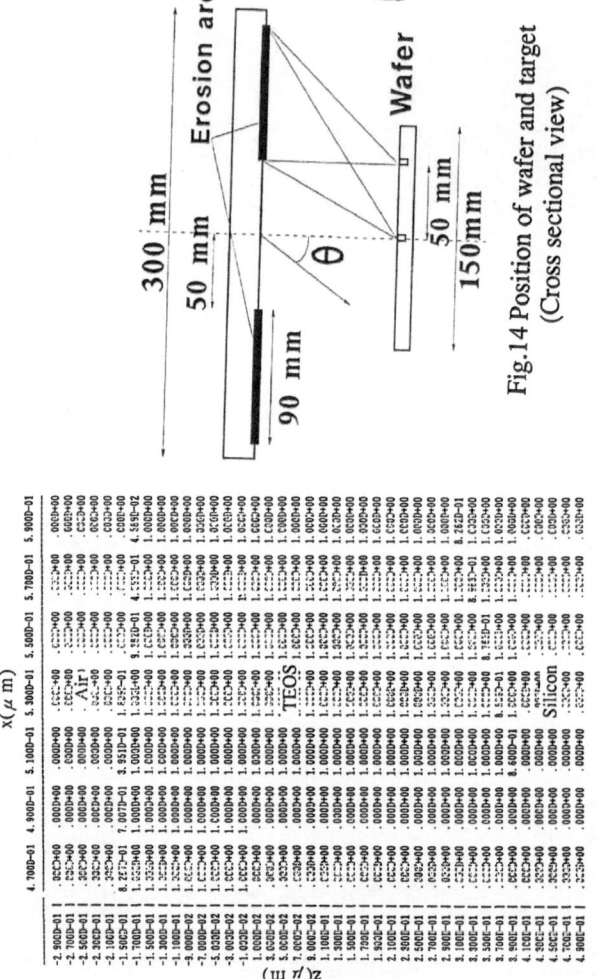

Fig.14 Position of wafer and target
(Cross sectional view)

Fig.13 Volume rate of TEOS

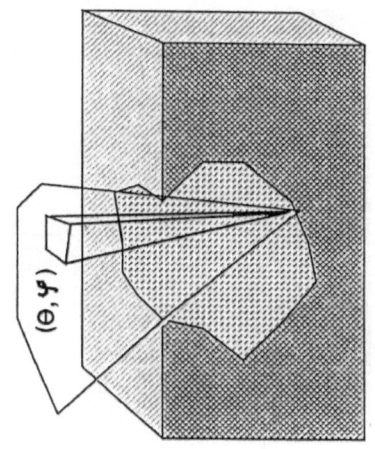

Fig.16 Surface shadowing

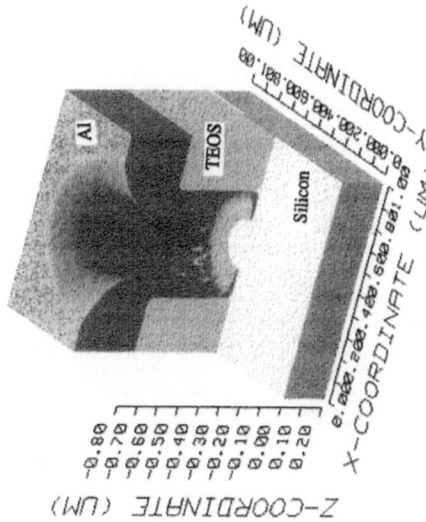

Fig.15 3D-simulation of aluminum sputter deposition

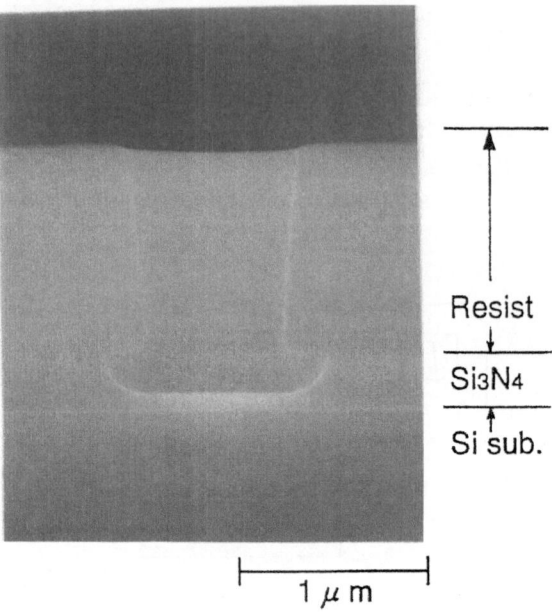

Fig.17 SEM photograph of isotropic dry etching(Si3N4)

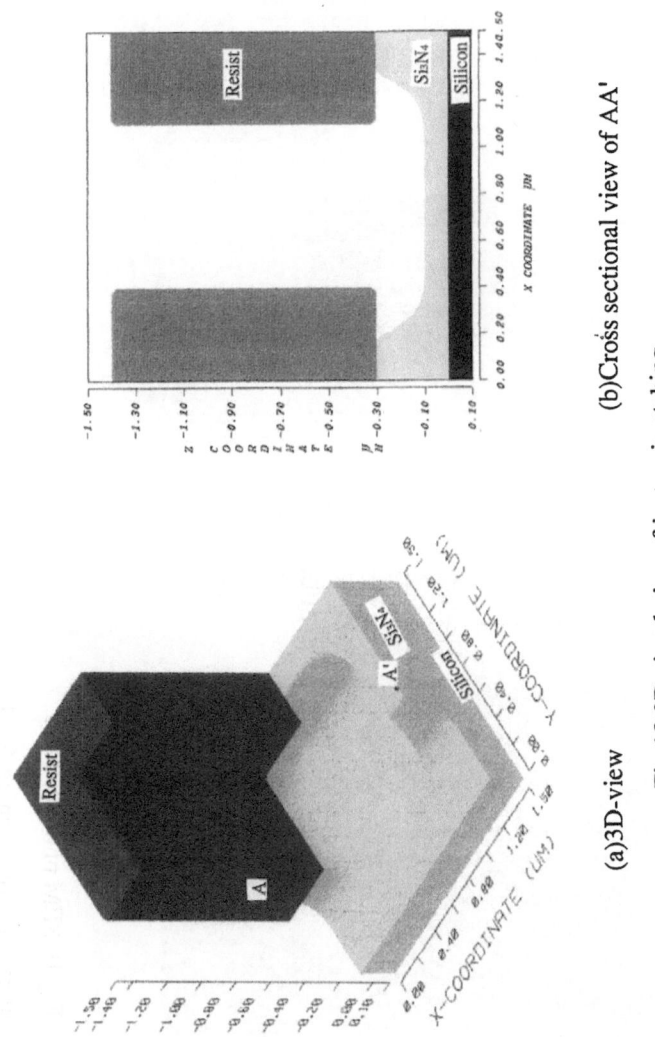

(b)Cross sectional view of AA'

(a)3D-view

Fig.18 3D-simulation of isotropic etching

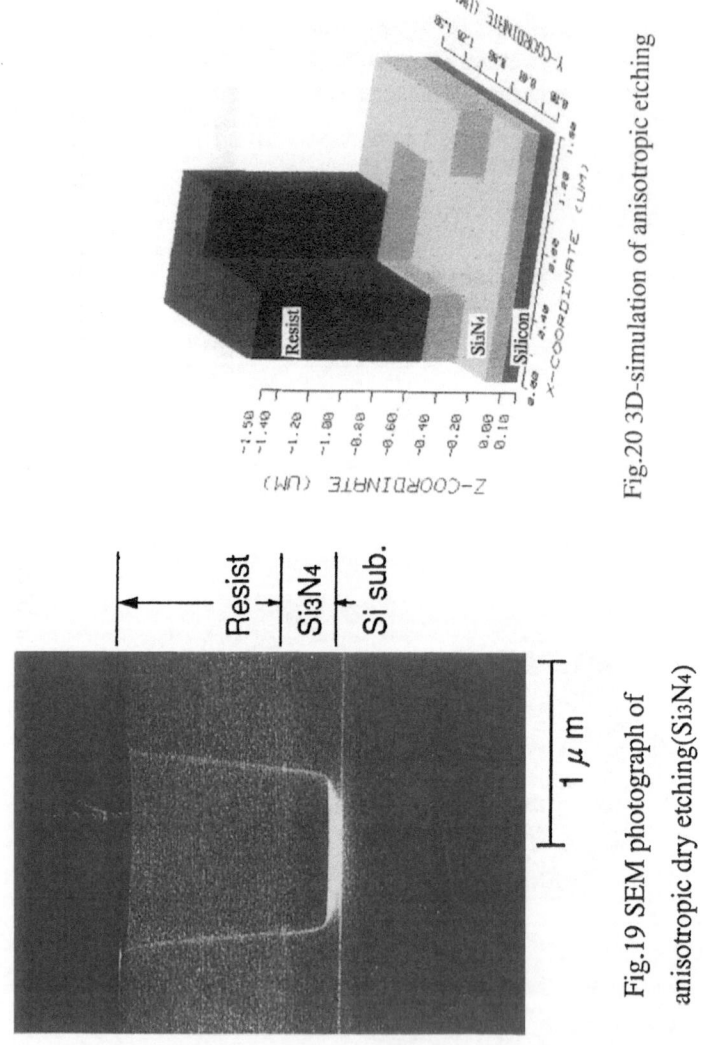

Fig.20 3D-simulation of anisotropic etching

Fig.19 SEM photograph of anisotropic dry etching(Si3N4)

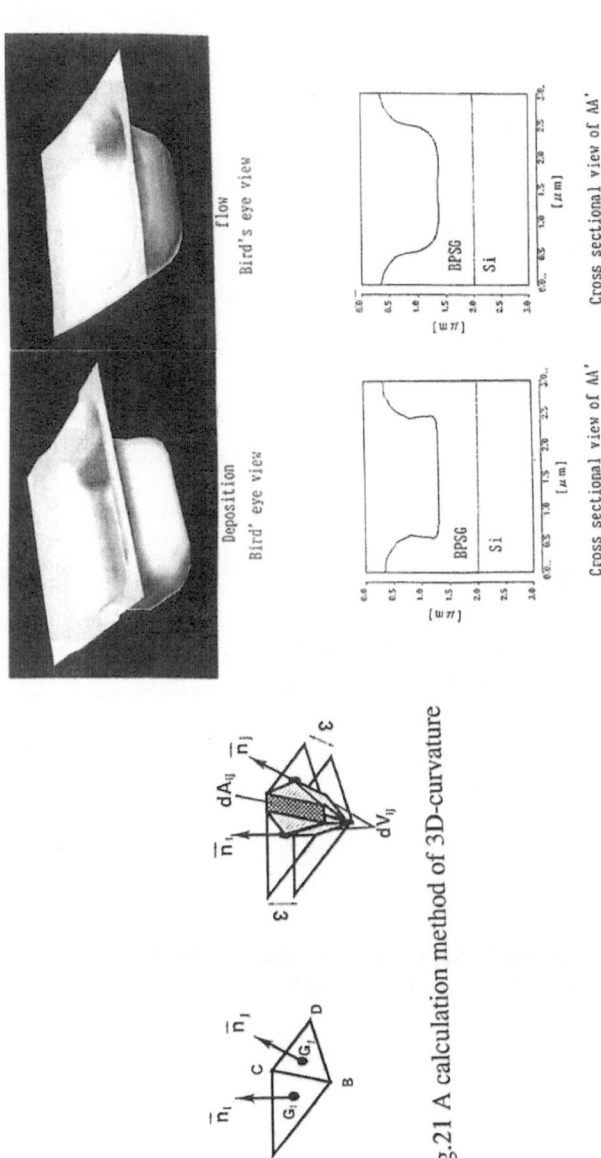

Fig.22 3D-simulation of BPSG flow

Fig.21 A calculation method of 3D-curvature

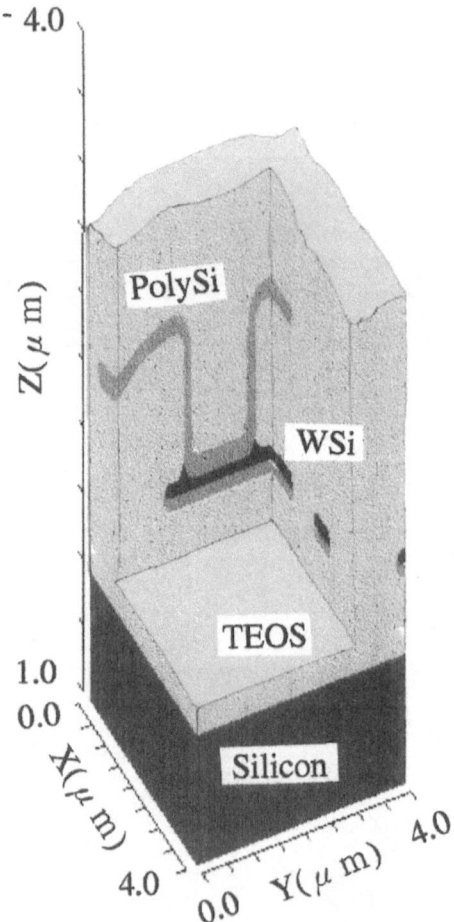

Fig.23 3D-simulation of sequential process steps
(4 μ m $\times$ 4 μ m $\times$ 5 μ m)

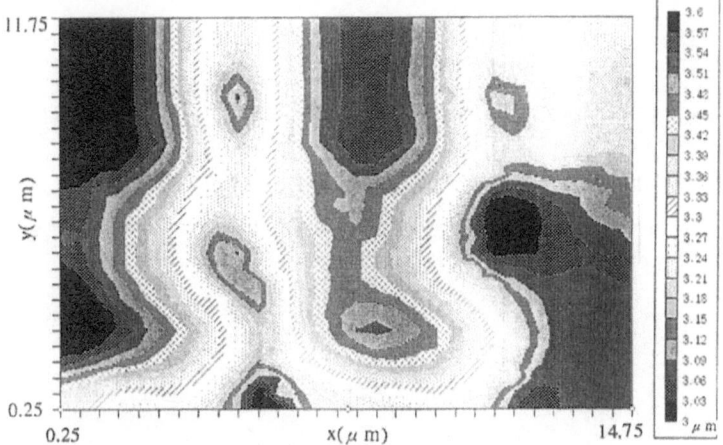

Fig.24 Contour of surface height in 3D-simulation result

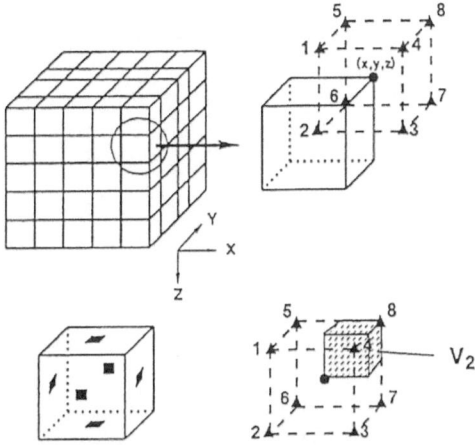

Fig.25 Division of Analysis region into cubic cells

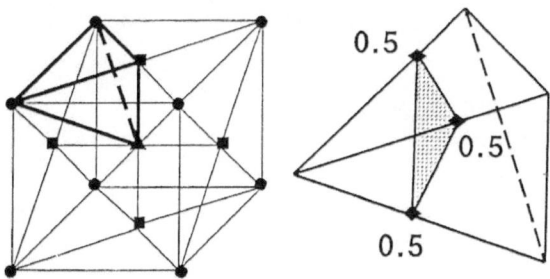

Fig.26 Division of a cubic cell into tetrahedrons

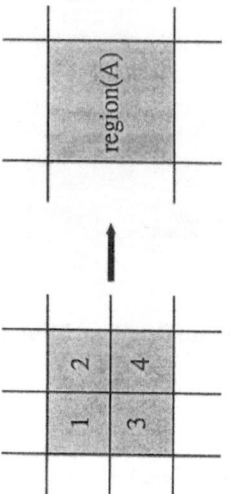

Fig.28 Transformation by renormalization

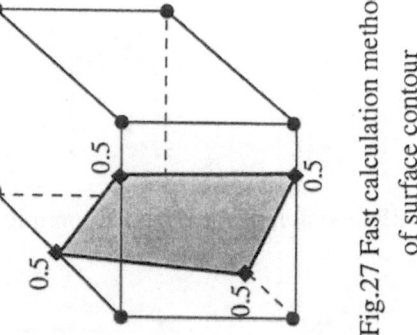

Fig.27 Fast calculation method of surface contour

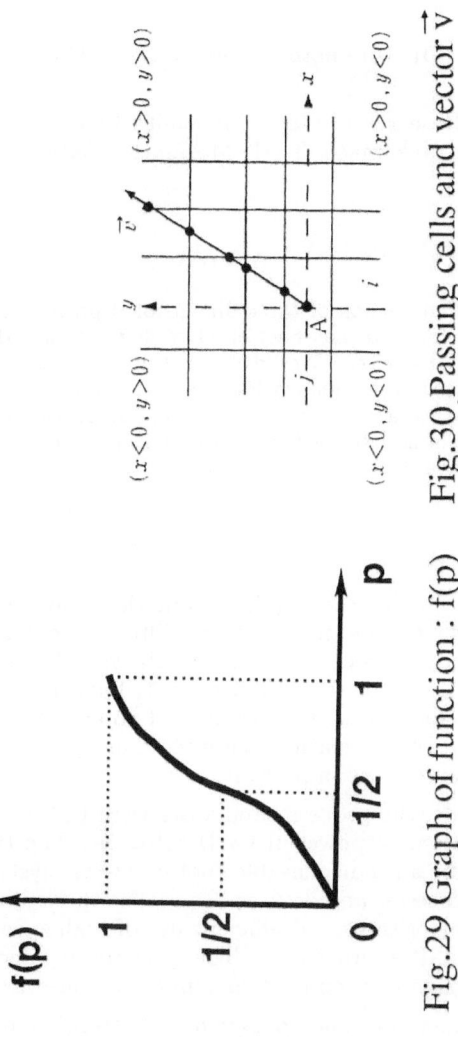

Fig.29 Graph of function : f(p)   Fig.30 Passing cells and vector $\vec{v}$

A Three-Dimensional Process Simulation using Advanced SMART-P program

H. Umimoto, S. Odanaka, A. Gohda

Semiconductor Research Center, Matsushita Electric Industrial Co., Ltd.
Yagumo-Nakamachi 3-1-1, Moriguchi, Osaka 570, Japan

Abstract

We developed a first version of three dimensional process simulator: SMART-P in 1987. This simulator has been used to develop the DRAM cells, CMOS processes and CCD devices. We advanced this program through these applications in some technical points, which are a TCAD system, physical models, numerical approaches and estimation techniques. In this paper, an advanced SMART-P program is described using three-dimensional simulation results of oxidation and BPSG flow.

1. Introduction

Miniaturization of VLSI devices requires numerical modeling of complicated fabrication processes such as ion-implantation, diffusion, oxidation and planarization processes. The numerical process simulation allows a better understanding of the multi-dimensional behavior of the impurity distribution, oxide growth and surface topography. These behaviors sometimes exhibit so-called three-dimensional effects. In scaling down of transistor, isolation and interconnect structures, three-dimensional process modeling becomes very important.

During the past decade, the process simulators were tightly integrated with the device modeling and become a powerful CAD software. The two-dimensional process simulators have become an indispensable tool in TCAD system for the development of VLSI fabrication processes and devices. To realize three-dimensional process CAD, numerical modeling, algorithms and efficient use of high-speed computers have been studied in the SMART-P approach [1]. The growing demand for TCAD is creating the need of TCAD design environment to support a wide variety of users [2].

Moreover, some authors continue to pay much attention to two-dimensional process modeling issues such as the transient enhanced diffusion and the stress effect for oxidation. Such physical phenomena strongly affect the electrical characteristics of sub-0.5μm devices. The physics-based approach is available for modeling of the complicated phenomena. In general, however, the physics-based approach requires extracting of many physical parameters for the accurate model and significantly increases computational cost. The rapid progress of the process development needs another approach using semi-empirical models with a database constructed by the experimental data.

For an accurate process modeling, it is further needed to make progress in the calibration technology for multi-dimensional dopant profiles and surface topography. So far, the SIMS measurement and cross-sectional SEM and TEM observations [3] are effectively used to estimate one-dimensional dopant profiles and two-dimensional surface topography, respectively. In the case of estimation for two-dimensional dopant profiles and three-dimensional surface topography, however, the further sophisticated method has to be developed to reveal the two- or three-dimensional effect more clearly.

We developed a first version of three-dimensional process simulator: SMART-P in 1987 [1], which is based on the finite difference approach to supercomputer Fujitsu VP-100 on the original operating system of Fujitsu. It has been used to develop the DRAM cells, CMOS processes and CCD devices [4, 5, 6]. We advanced this program through these applications in some points of view, which are a TCAD system, physical models, numerical approaches and estimation techniques [7, 8, 9, 10, 11].

This paper is an extensive overview of three-dimensional simulations using advanced SMART-P program, which is reconstructed on the UNIX operating system. In Section 2, a TCAD system in Matsushita. is briefly introduced. In Section 3, numerical process modeling in the Si/SiO_2 system is described. In Section 4, it is shown that advanced SMART-P program realized this three-dimensional process modeling using efficient numerical algorithms. Finally, Section 5 demonstrates the three-dimensional simulations of LOCOS process with the stress effect and BPSG flow with the two typical mask structures. These calculated results are estimated by the experimental data obtained by using SEM and STM observations.

2. SMART-P in TCAD System

In this section, a SMART TCAD system in Matsushita is briefly described.

2.1. System Structure

The SMART TCAD system has been reconstructed on the UNIX operating system which realizes the distributed environment using many workstations: Panasonic P-4000 and supercomputer: Fujitsu VPX-220. The workstations and a supercomputer are connected using FNS on the ethernet. The structure of SMART TCAD system is outlined in Fig.1. The SMART system consists of the 3D process simulator: SMART-P, a 3D device simulator: SMART [12] on the vector type supercomputer, a task manager, the 3D color graphic tool:Views-II and SIMS database on workstations.

2.2. Task Manager

The use of the UNIX operating system allows the handling of some tasks for simulations using the SMART system. The task manager: SMART-TM is written by the Perl which likes as C, sed, awk, and shell programming languages for generating inputdecks of simulator. This command-based task manager supervises and controlls the simulation tasks.

The SMART-TM has some libraries which are template cards of process flows, device structures including grid information, bias conditions and model descriptions. It generates a input deck of SMART-P and SMART by combination with some template cards in such libraries. Users can define the description for process and device design in the template card easily.

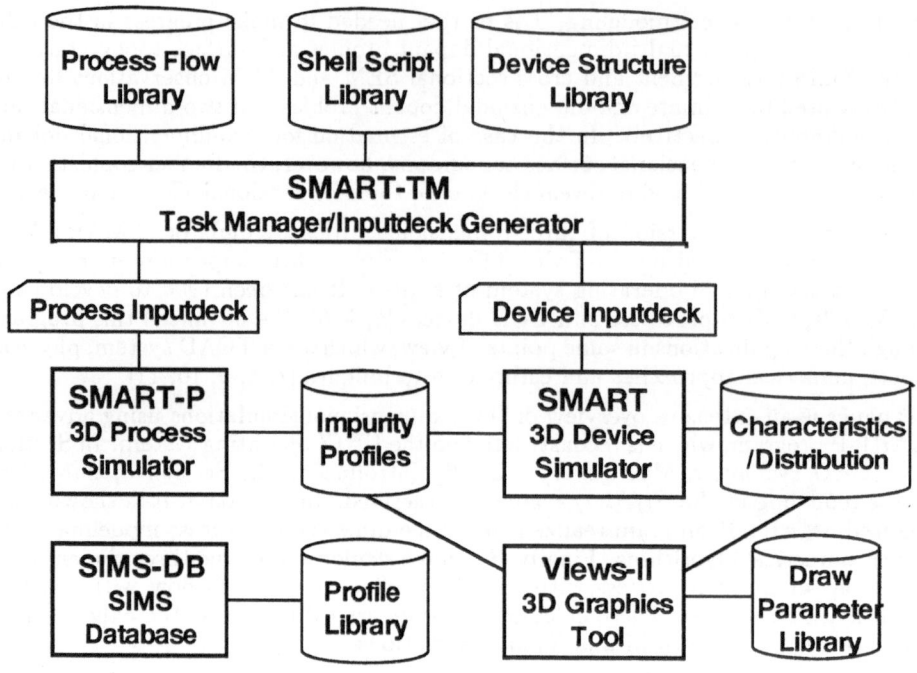

Figure 1: Structure of SMART TCAD system.

The SMART-TM also has the shell libraries which repesent some simulation tasks such as the threshold voltage lowering as a function of the gate length. The simulation tasks are described in Perl script.

The task manager also has a filter program to connect the process simulator with the device simulator automatically.

2.3. Three-Dimensional Color Graphic Tool

The 3D graphic system is an important component in the 3D TCAD system to understand the simulation results easily and effectively. Views-II is a in-house 3D color graphic system based on PEX implemented in X11R5. Table 1 shows the software structure of Views-II. The menu panel is built by using the X-Toolkit, and the draw

Views-II	
MENU	DRAW/CONTROL
XToolkit	*PHIGS/PHIGS PLUS*
X11R5/PEX-SI	
UNIX	

Table 1: Software structure of Views-II

and control functions are realized by using Phigs functions. It provides a good user interface operating like a pop up menu shown in Fig.2. Views-II reads the result files

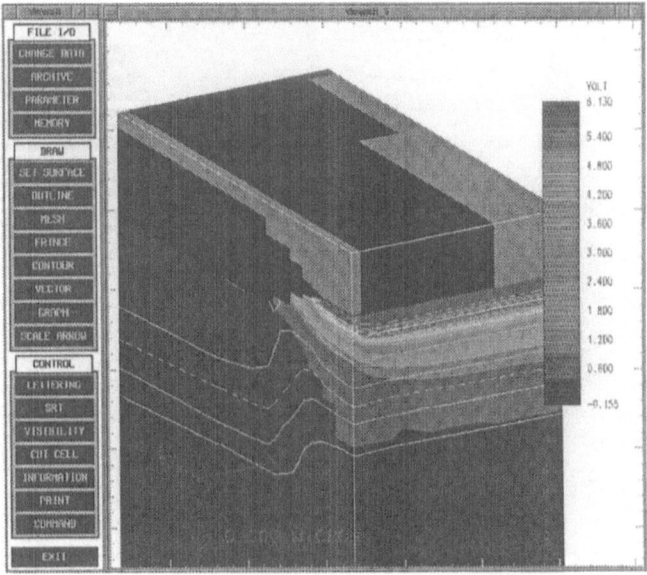

Figure 2: Window image of Views-II

generated by SMART and SMART-P. Scalar and Vector data in the result files are visualized by the following functions in Views-II.

- Draw Function
  - Contour Plots
  - Vector Plots
  - Mesh Plots
  - 1-D Cross-Sectional Plots
- Control Function
  - Cutting Plane
  - Scaling, Rotating and Translating
  - Stacking of Draw and Control Parameter
  - Memorizing of Picture

2.4. SIMS database

A SIMS database system: SIMS-DB has been developed to store and reuse many measured profiles. This program is used to compare the experimental data with the simulated results, and to extract the parameter of implant statistics. It is one of tools which realize accurate predictions on the basis of the semi-empirical models. Fig.3 shows the window image of SIMS-DB. SIMS-DB is built on a relational database. It provides an good user's interface that allows users to search profiles through the

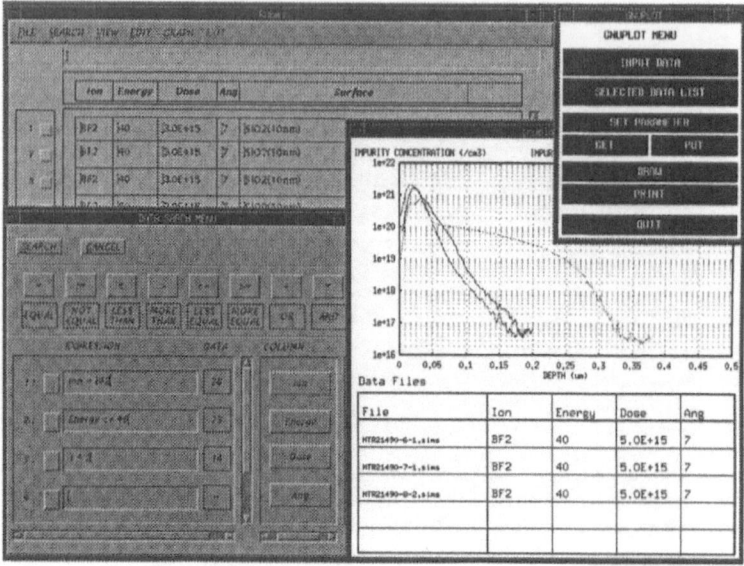

Figure 3: Window image of SIMS database system

database by choosing specified criteria. The criteria is ion species, dose, energy, annealing time and temperature, and device name. SIMS-DB includes SIMS libraries which are obtained in the development of DRAM cell, CMOS process and CCD devices in Matsushita. The data can be viewed by using the spread sheet and graphics following the specified criteria.

3. Physical Model

In this section, the advanced physical models in SMART-P are described at each process.

3.1. Implantation

The original SMART-P utilizes analytical models for the simulation of ion implantation. Vertical ion profiles are described by using joined half-Gaussian distributions for arsenic and phosphorus and by using Pearson IV distribution for boron. The range profiles for multi Si/SiO_2 layer with mask layers are taken from [13]. The lateral convolution with Gauusian distribution is employed to obtain the lateral distribution in spatial three dimensions. Moreover, the three-dimensional profiles of the large tilt angle implanted ions [14] can be calculated by using the following rotaion transformation technique.

$$x' = x \cos\theta \cos\phi + y \sin\theta \cos\phi + z \sin\phi \qquad (1)$$
$$y' = -x \sin\theta + y \cos\theta \qquad (2)$$
$$z' = -x \cos\theta \sin\phi - y \sin\theta \sin\phi + z \cos\phi \qquad (3)$$

where ϕ and θ are tilt and rotation angles of incident ion in the coordinate system (x, y, z) as shown in Fig.9, respectively. In the coordinate system (x', y', z'), the incident ion is parallel to the z' axis.

To obtain range profiles for implanted ions, SMART-P utilizes lookup tables based on LSS theory [15]. However, there are some discrepancy between the values of range tables based on LSS theory and experimental data because of channeling of ion. The advanced SMART-P is used for range data extracted by measured profiles in standard CMOS processes, which are stocked by the SIMS database system.

3.2. Diffusion

Impurity diffusion is one of the key steps for simulating fabrication processes of VLSI devices. The redistribution of impurities in the Si/SiO_2 system is modeled by solving the following equation for the jth impurity.

$$\frac{\partial C_{jSi}}{\partial t} = \nabla D_{jSi}(f_{C_j}\nabla C_{jSi} \pm f_{E_j}\nabla C_{net}) \qquad in\ silicon \qquad (4)$$

$$\frac{\partial C_{jox}}{\partial t} = D_{jox}\nabla^2 C_{jox} \qquad\qquad\qquad in\ oxide \qquad (5)$$

where C_{jSi} and D_{jSi} are the impurity concentration and diffusivity in silicon. C_{jox} and D_{jox} is the impurity concentration and diffusivity in oxide. C_{net} = donors - acceptors. f_{C_j} and f_{E_j} are the reduction factor due to clustering and the electric field enhancement factor for impurity concentration, receptively. At the silicon-oxide interface, impurity concentrations are adjusted in accordance with the particle flux effects of moving boundary and impurity segregation:

$$D_{jox}\frac{\partial C_{jox}}{\partial n} = -h_j(C_{jox} - \frac{1}{m_j}C_{jSi}) \qquad on\ silicon/oxide\ interface \quad (6)$$

$$D_{jSi}\frac{\partial C_{jSi}}{\partial n} = D_{jox}\frac{\partial C_{jox}}{\partial n} + \frac{v_{ox}\cdot n}{\alpha}(C_{jox} + \alpha C_{jSi}) \ on\ silicon/oxide\ interface (7)$$

where h_j is the mass transfer coefficient, m_j is the segregation coefficient, and v_{ox} is the velocity of the silicon/oxide interface. The quantity α is the volume ratio of consumed silicon to grown oxide and n is the unit vector normal to the silicon/oxide interface. It is known that diffusion in silicon is induced by both vacancy- and interstitial-assisted diffusion mechanisms. To solve the accurate impurity distribution, the modeling of coupled diffusion of impurities and point defects under interaction is necessary, in particular, for the OED (Oxidation Enhanced Diffusion) under oxidizing conditions and TED (Transient Enhanced Diffusion) after ion-implantation at the low diffusion temperature. The advanced SMART-P adopts a unique approach for three-dimensional diffusion modeling. The OED model considers the local distribution of the effective interstitial [16], and for TED a simple semi-empirical model is developed without considering the local distribution of point defects [17]. The enhanced diffusivity is modeled by using the following expression.

$$D = D_N + \Delta D_{OED} + \Delta D_{TED} \qquad\qquad (8)$$

where D_N is the effective vacancy-related diffusion coefficient including the high concentration effect of impurity diffusion under nonoxidizing conditions, ΔD_{OED} and ΔD_{TED} are the the diffusivity components enhanced by OED and TED, respectively.

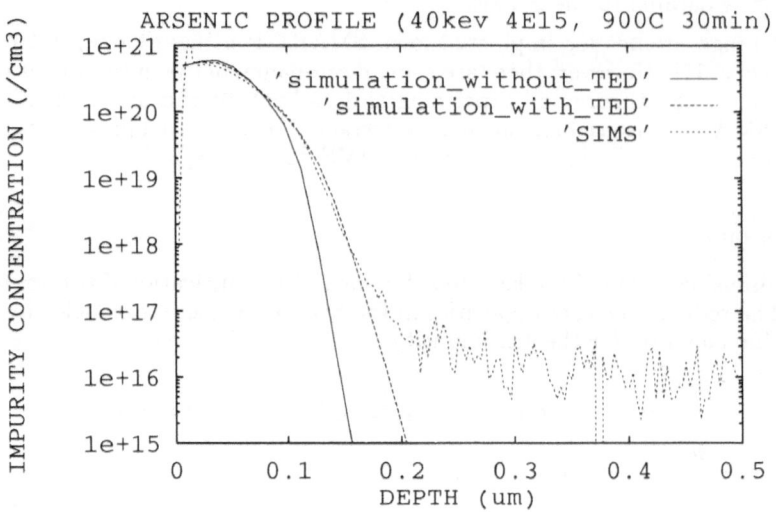

Figure 4: Drain profile of arsenic for n-channel MOSFET

These components of diffusivity are expressed as

$$\Delta D_{OED} \;=\; f_I \cdot D_i \cdot \left(\frac{C_I - C_I^{eq}}{C_I^{eq}} \right) \tag{9}$$

$$\Delta D_{TED} \;=\; K_{TED} \cdot D_i \exp\left\{ \frac{-t}{\tau} \right\} \tag{10}$$

where D_i is the intrinsic diffusivity and C_I is the silicon interstitial concentration and C_I^{eq} is the equilibrium interstitial concentration and f_I is the interstitial component of diffusion. The parameters of K_{TED} and τ in the TED model are extracted by some measured impurity profiles in the standard CMOS process such as a well, channel, and drain formation. These measured profiles are accumulated by the SIMS database system. SMART-P dose not take account of the local distribution of point defects induced by ion implantation. Fig.4 and Fig.5 show the drain profiles obtained by the SIMS and simulation for n-channel and p-channel MOSFET, respectively. The semi-empirical TED model is effective to calculate accurate profiles in the depth direction.
  On the other hand, the OED model parameters C_I^{eq} and f_I are taken from references [18, 19] and the effective interstitial distribution in silicon is numerically calculated in three dimensions to solve the following diffusion equation by using the effective diffusivity of interstitial D_{Ieff} [18].

$$\frac{\partial C_I}{\partial t} = D_{Ieff} \nabla^2 C_I \qquad in \quad silicon \tag{11}$$

$$D_{Ieff} = 8.6 \times 10^5 \exp\left\{ \frac{-4.0eV}{k_B T} \right\} \; . \tag{12}$$

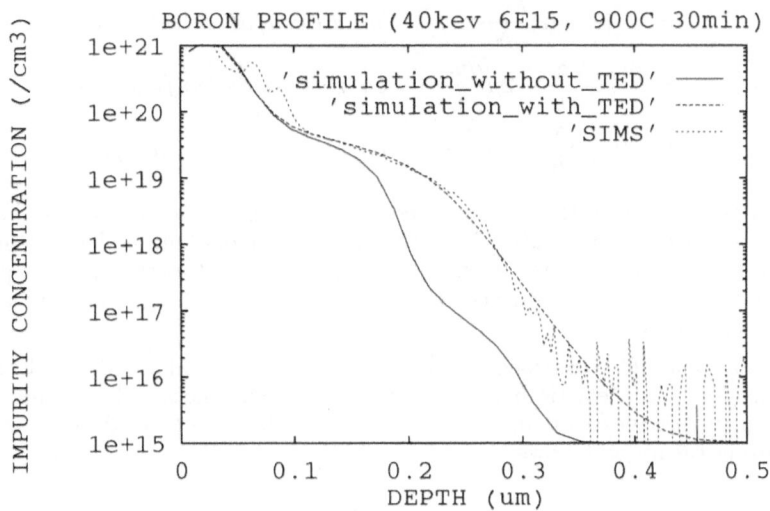

Figure 5: Drain profile of boron for p-channel MOSFET

For the effective interstitial distribution, the supersaturation model of interstitials under oxidizing condition is assumed at the silicon surface in the Si/SiO2 system [16].

$$- D_{Ieff}\frac{\partial C_I}{\partial n} = K_{gen} \exp\left\{\frac{Q_{gen}}{k_B T}\right\} \left(\frac{dT_{ox}}{dt}\right)^n + K'_{gen} - K_I(C_I - C_I^{eq}) . \qquad (13)$$

At the bottom of silicon, the boundary condition is given by $C_I = C_I^{eq}$.

3.3. Oxidation

The physical model of oxidation is based on a steady-state oxidant diffusion and a slow incompressible viscous flow of oxide [20]. In the advanced SMART-P program, this oxidation model is extended to include the stress effects in three dimensions [7, 11]. The oxidant diffusion is written as

$$\nabla \cdot (D_{eff}\nabla C) = 0 \qquad in \; oxide \qquad (14)$$

where C is the oxidant concentration and D_{eff} is the effective stress-dependent diffusivity of oxidant. The boundary conditions follow the Deal-Grove one-dimensional oxidation model [21]. The slow incompressible viscous flow of oxide is expressed by the following hydrodynamic equation with zero divergence:

$$\mu\nabla^2 v \;=\; \nabla P \qquad in \; oxide \qquad (15)$$
$$\nabla \cdot v \;=\; 0 \qquad in \; oxide \qquad (16)$$

where μ, v, and P are the stress dependent viscosity of oxide, the velocity vector of an oxide element, and pressure (positive sign for compression), respectively. The following boundary conditions for equation (15) are used.

$$v = -(1-\alpha)\frac{k_S C}{N}n \qquad on\ silicon/oxide\ interface \qquad (17)$$

$$P = P_a - \frac{\gamma}{R} + 2\mu\frac{\partial v_n}{\partial n} \qquad on\ free\ oxide\ surface \qquad (18)$$

$$P = P_f - \frac{\gamma}{R} + 2\mu\frac{\partial v_n}{\partial n} \qquad on\ nitride/oxide\ interface \qquad (19)$$

where α is the volume ratio of consumed silicon to grown oxide, k_S is the stress-dependent surface reaction rate, N is the number of oxidant molecules in a unit oxide volume, n is a unit vector normal to the boundary, P_a is the ambient pressure, P_f is the nitride stress, γ is surface tension, R is local curvature, and v_n is normal component of velocity.

In the conventional LOCOS process [22], the three-dimensional nitride mask structure can be considered as a elastic beam with a constant thickness. The nitride bending stress P_f on the nitride/oxide interface is simply modeled in three dimensions by the beam bending theory as follows:

$$P_f = E_{nit} I \Delta \frac{1}{R} \qquad (20)$$

$$I = \frac{D_{nit}^3}{12(1-\nu)^2} \qquad (21)$$

where E_{nit} is the Young's modulus of nitride, I is the moment of inertia, $T(x,y,t)$ is the surface position of BPSG, D_{nit} is the nitride thickness and ν is the Poisson ratio of nitride.

In nonplanar oxidation process, the oxide growth depends on the oxidation-induced stress. The advanced SMART-P program takes into account such stress effects in the oxidation model. This is expressed by including the following stress-dependent physical parameters into oxidation kinetic equations (14) and (15), and boundary conditions (17), (18) and (19)[23].

$$k_S = k_0 \exp\left\{\frac{\sigma_{nn} V_k}{k_B T}\right\} \qquad (22)$$

$$D_{eff} = D_0 \exp\left\{-\frac{P V_D}{k_B T}\right\} \qquad (23)$$

$$\mu = \mu_0 \exp\{\beta P\}, \quad for\ P < 0 \qquad (24)$$

here σ_{nn} is the normal stress (negative sign for compression) at the silicon/oxide interface, V_k is the reaction jump-volume in silicon to oxide, V_D is the active diffusion volume, and β is an empirical parameter. The zero-stress parameters k_0 and D_0 are derived from the experimental linear and parabolic rate constants for planar silicon oxidation. The orientation dependence of k_0 is taken into account by the linear interpolation from the value of k_0 on $< 100 >$, $< 110 >$ and $< 111 >$ oriented surface.

3.4. BPSG Flow

The planarization process utilizing the BPSG flow is widely used to develop multi-level wiring and stack capacitor cells [24]. The capability of this planarization process

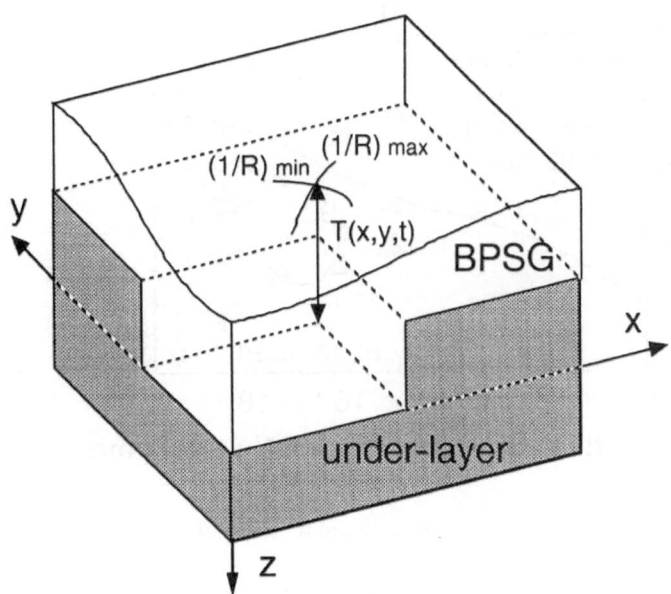

Figure 6: Definition of the surface curvature in three-dimensional space.

strongly depends on the temperature and impurity concentration of BPSG [26]. In advanced SMART-P program, a BPSG flow model is newly implemented. The physical model of BPSG flow is also based on a slow incompressible viscous flow of BPSG as well as oxide [25]. It is assumed that the driving force of the BPSG flow is a surface tension on the free BPSG surface and the BPSG flow does not slip on the BPSG/under-layer interface. This yields the following boundary conditions:

$$P = P_a - \frac{\gamma}{R} + 2\mu\frac{\partial v_n}{\partial n} \qquad on\ free\ BPSG\ surface \qquad (25)$$

$$v = 0 \qquad\qquad on\ BPSG/underlayer\ interface. \qquad (26)$$

The BPSG flow is driven by the surface tension and hence the boundary condition (25) at the free surface of BPSG includes the numerical modeling of the surface curvature. Fig.6 shows a schematic hole structure after flow anneal. The curvature $\frac{1}{R}$ of the free surface in three dimensions is modeled by using a two times mean curvature $2H$, which corresponds to a summation of the maximum normal curvature $(\frac{1}{R})_{max}$ and the minimum normal curvatures $(\frac{1}{R})_{min}$:

$$\frac{1}{R} = 2H = \left(\frac{1}{R}\right)_{max} + \left(\frac{1}{R}\right)_{min} \qquad (27)$$

Assuming that there is no overhang structure on the free surface of BPSG, the three-dimensional surface curvature is expressed by using surface position $T(x, y, t)$ as fol-

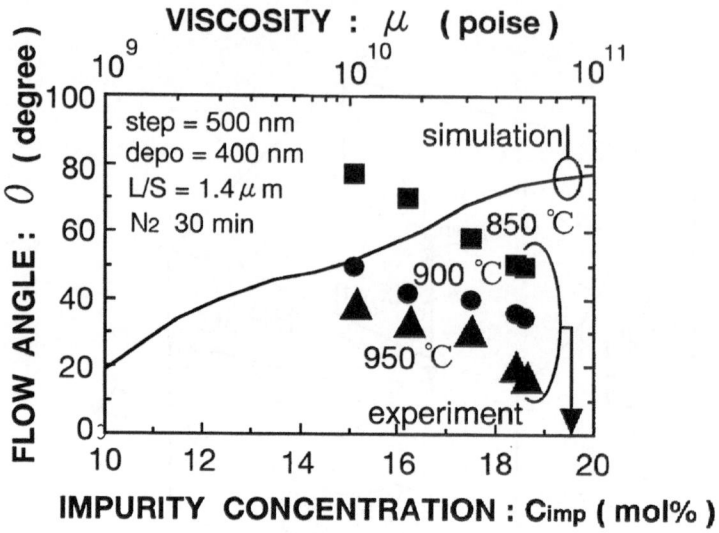

Figure 7: The flow angle versus impurity concentration of B_2O_3 and P_2O_5 in BPSG and viscosity.

lows:

$$\frac{1}{R} = \frac{\left\{1 + \left(\frac{\partial T(x,y,t)}{\partial x}\right)^2\right\} \frac{\partial^2 T(x,y,t)}{\partial y^2} - 2\frac{\partial T(x,y,t)}{\partial x}\frac{\partial T(x,y,t)}{\partial y}\frac{\partial^2 T(x,y,t)}{\partial x \partial y} + \left\{1 + \left(\frac{\partial T(x,y,t)}{\partial y}\right)^2\right\} \frac{\partial^2 T(x,y,t)}{\partial x^2}}{\left\{1 + \left(\frac{\partial T(x,y,t)}{\partial x}\right)^2 + \left(\frac{\partial T(x,y,t)}{\partial y}\right)^2\right\}^{\frac{3}{2}}}$$

(28)

The viscosity μ of BPSG is newly modeled as a function of the temperature and impurity concentration in BPSG. It is well known that the BPSG flow strongly depends on the concentration of B_2O_3 and P_2O_5 in BPSG [26]. It is reported that the flow angle θ, which is the maximum tangential angle of the BPSG surface, is expressed as a function of the total concentration of B_2O_3 and P_2O_5 in BPSG [27]. This report implies that the viscosity of BPSG under the various flow temperature and total impurity concentration can be extracted from a comparison of the flow angle θ between experiments and simulations in the line structure with the equi-line. Fig.7 shows the measured flow angle versus total impurity concentration C_{imp} of B_2O_3 and P_2O_5 in BPSG at 850°C, 900°C and 950°C. Also, as shown in Fig.7, the calculated flow angle is indicated as a function of the viscosity μ. Calculating the same flow angle as the experimental data, the viscosity versus impurity concentration characteristics are obtained at each temperature as shown in Fig.8. In a nitrogen ambient, the flow temperature and impurity concentration dependent viscosity model is obtained as follows.

$$\mu(T, C_{imp}) = \mu_0(T) \exp\left\{-k(T)C_{imp}\right\}$$

(29)

$$\mu_0(T) = 9.184 \times 10^{-18} \exp\left\{\frac{6.921 eV}{k_B T}\right\}$$

(30)

$$k(T) = -7.8 \times 10^{-6} T^2 + 1.643 \times 10^{-2} T - 8.116$$

(31)

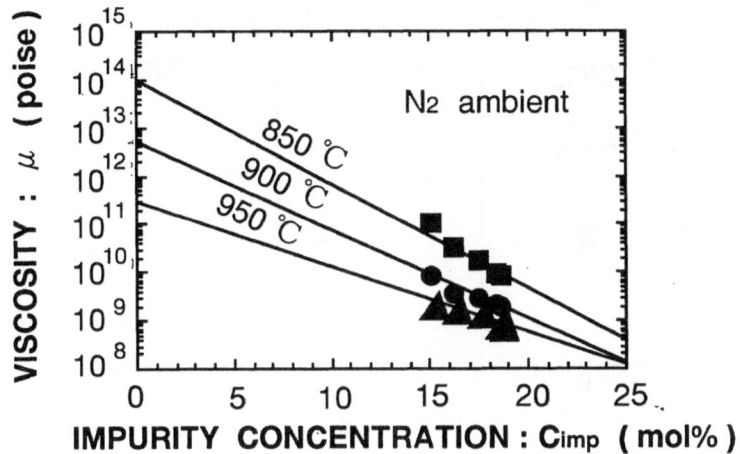

Figure 8: Viscosity of BPSG versus impurity concentration.

4. Efficient Numerical Algorithms

To realize the numerical modeling of the nonplanar Si/SiO_2 structure, the efficient numerical algorithms are needed in three dimensions. In this section, the coordinate transformation method, relaxation technique and matrix solver are described.

4.1. Coordinate Transformation Method

Numerical modeling for the three-dimensional nonplanar Si/SiO_2 structures is based on a finite difference approach using the coordinate transformation method [28]. It is expressed as

$$\xi = x \tag{32}$$

$$\eta = y \tag{33}$$

$$\zeta = \frac{z - T(x,y,t)}{H(x,y,t) - T(x,y,t)} \tag{34}$$

$$\zeta = \frac{z - H(x,y,t)}{B - H(x,y,t)} \tag{35}$$

$$\tau = t \tag{36}$$

where $T(x,y,t)$ is the top surface of the oxide and $H(x,y,t)$ is the silicon/oxide interface, and B is the initial position of the silicon bottom. Fig.9 shows schematic cross-sections of the grid system before the coordinate transformation and after the coordinate transformation. By using this coordinate transformation method, the physical Si/SiO_2 structure in three spatial dimensions is transformed into two computational rectangular regions of oxide and silicon shown in Fig.9. The diffusion and viscous flow kinetic equations are also transformed into the coordinate system (ξ,η,ζ,τ). This method allows the moving boundary modeling. In addition, this method can make it easy to handle the deformation of mesh, which is a key issue in the finite element method at updating boundary.

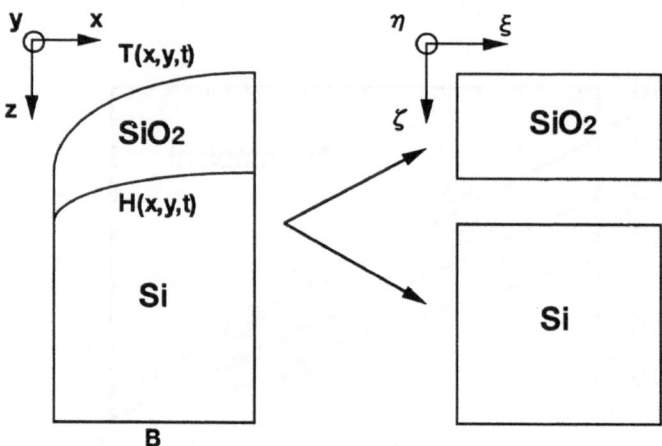

Figure 9: Coordinate transformation method.

4.2. Relaxation Technique

We have already developed an explicit calculation procedure for oxidation kinetic equations without the stress effect in the original SMART-P. It requires no numerical iteration for velocity/pressure calculation procedure. In advanced SMART-P program, this explicit calculation procedure is also applied to solve the BPSG flow equation. Moreover, to obtain the self-consistent solution of oxidation kinetic equations with the stress effect, a relaxation technique is applied to the oxidant and velocity/pressure calculation procedure as shown in Fig.10. After solving oxidation kinetic equations, the normal stress and pressure are relaxed by using two relaxation factors ω_s and ω_p, respectively

$$\sigma_{nn}^m = \omega_s \sigma_{nn}^m + (1 - \omega_s)\sigma_{nn}^{m-1} \qquad (37)$$
$$P^m = \omega_p P^m + (1 - \omega_p)P^{m-1} \qquad (38)$$

where m represents the m-th iteration at the each time step. The stress-dependent physical parameters (22)-(24) are calculated by using these relaxed stress values. The iteration procedure shown in Fig.10 is carried out until the normal stress and pressure are converged.

4.3. Matrix Solver

Another approach to achieving high-speed computation is the efficient use of a vector processor. In order to solve the coupled nonlinear diffusion equation (4) in silicon and the diffusion equation (5) in oxide, SMART-P uses the RRK (Rational Runge-Kutta) method for the time integration and the upwind scheme for spatial discretization of the convection terms. In the advanced SMART-P program, the asymmetric matrix solver for the oxidation kinetics equation (14)(15) and the effective interstitial diffusion equation (11) is newly developed by using the BiCGSTAB method with the incomplete LU decomposition (Incomplete LU Biconjugate Gradients STAB Method: ILUBiCGSTAB). It is coded in vectorized manner using the list vector method on the basis of the hyperplane ordering which avoids recurrence in inner DO loops in

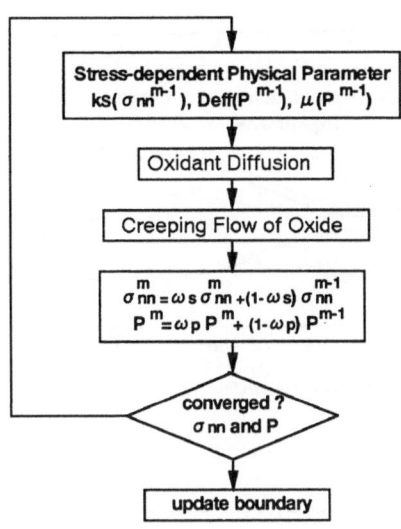

Figure 10: Flowchart of relaxation technique.

a nest of DO loops. The computational time becomes more attractive as shown in Table. 2.

Equations	Total Grid	CPU Time for Matrix Solver		Accelerated Rate
		BiCG (sec)	STAB (sec)	BiCG/STAB
Oxidation Kinetics	41472	383	247	1.55
Interstitial Diffusion	196992	1613	842	1.92

Table 2: Comparison of cpu time between ILUBiCG and ILUBiCGSTAB

5. Three-Dimensional Simulation and It's Estimation

5.1. LOCOS Simulation

5.1.1. Nitride Bending Stress

To demonstrate the stress effects induced by the nitride mask bending, the calculated oxide shapes in the line structure are compared with the experimental data as a function of the nitride thickness D_{nit}. Fig.11 (a)-(c) shows cross sectional oxide shapes in the line structure with the $400\,\mathring{A}$, $1200\,\mathring{A}$ and $1800\,\mathring{A}$ thick nitride masks. The experimental data is obtained by the cross-sectional SEM observation. In the case of the thin nitride mask ($400\,\mathring{A}$) as shown Fig.11 (a), both calculated oxide shapes are in a good agreement with the experimental data. In the case of the thick nitride mask ($1200\,\mathring{A}$ and $1800\,\mathring{A}$) as shown Fig.11 (b) and (c), the oxide is suppressed at the mask edge in both experiment and simulations. However, there is a marked discrepancy between the calculated results without the stress effect and the experimental data

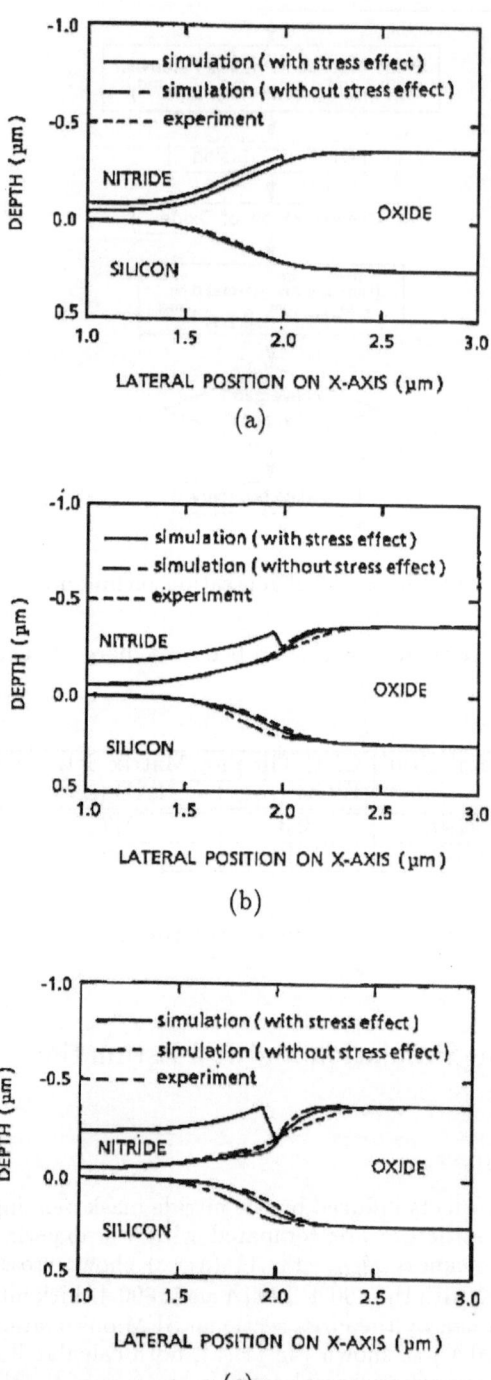

Figure 11: Cross-sectional LOCOS shape in the line structure.

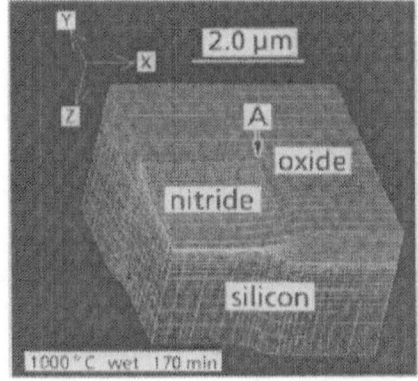

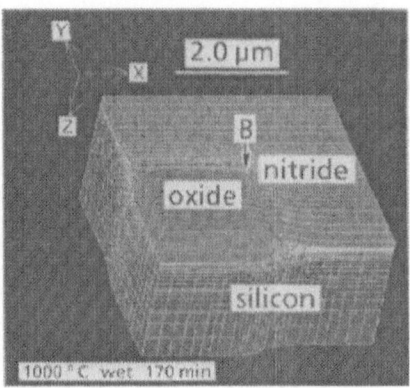

(a) HOLE STRUCTURE   (b) ISLAND STRUCTURE

Figure 12: Three-dimensional LOCOS structure obtained by simulation in the hole and island structure.

at the position of the silicon/oxide interface below the nitride mask edge. On the other hand, the oxide shape calculated with the stress effect gives a better agreement with the experimental data. This is because the compressive pressure induced by the nitride mask bending causes the retardation of the oxide growth below the nitride mask edge [7]. The results show that the beam bending equation (20) with a constant thickness D_{nit}. is well modeled for the nitride mask bending stress P_{nit}.

5.1.2. Mask Structure Effect on Bird's Beak Length

Two typical structures of LOCOS were investigated to understand three-dimensional effects of oxidation. Fig.12 (a)-(b) shows the calculated oxide shapes for such typical structures which are named as (a) hole structure (it models the contact structure) and (b) island structure. In Fig.12, the calculated oxide thickness is larger at the point A (hole structure) than that at the point B (island structure).

To clarify the mask structure effects on the oxide growth, the lateral expansion of oxide underneath the mask is investigated by comparing the top view SEM photographs with the calculated results. Fig.13 (a)-(b) shows the top view SEM photographs and simulation results near the corner of the mask edge. The experimental bird's beak edge is observed as the interface line between the oxide region and the silicon region after etching of the pad oxide. The bird's beak edge for simulations is determined as the position where the oxide thickness is $1000 \mathring{A}$. The calculated position of the bird's beak edge gives a good agreement with the experimental data in the hole and island structures. In particular, at corners of the mask edge, the rounding of the bird's beak edge is simulated well in both structures. The bird's beak length at the corner of the mask edge is much enhanced in the hole structure and it is retarded in the island structure.

The difference of the bird's beak length between the hole and island structures can be explained by the following two reasons. The first is the higher oxidant concentration at

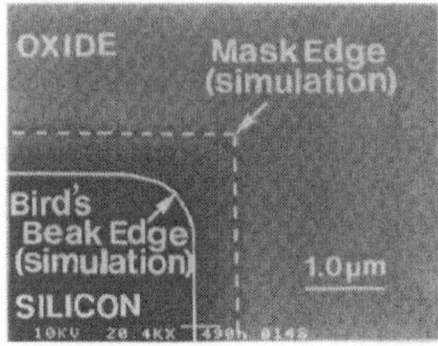 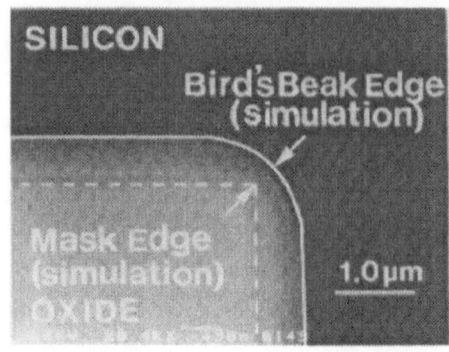

(a) HOLE STRUCTURE   (b) ISLAND STRUCTURE

Figure 13: Top-view of LOCOS structure.

the corner of the mask edge in the hole structure and the lower oxidant concentration in the island structure. Fig.14 (a)-(b) shows the distributions of the calculated oxidant concentration on the silicon/oxide interface in the hole and island structures. It is found that the oxidant concentration at the corner of the mask edge is higher in the hole structure and it is lower in the island structure. This phenomena is induced by the different behavior of the oxidant flux between hole and island structures. In the hole structure, the oxidant flux around the mask region concentrates to the center of the mask region. In the island structure, the oxidant flux from the mask window spreads into the mask region. As a result, the oxide growth rate under the corner of mask edge is enhanced in the hole structure and it is retarded in the island structure.

The second reason is shown in Fig.15. In the island structure, the compressive pressure caused by the nitride mask bending is larger than that in the hole structure. At the corner of the mask edge, the compressive pressure is larger in the island structure than that in the hole structure. This is because the masked area near the corner is larger in the island structure than that in the hole structure. This stress induces the retardation of the oxide growth, especially at the mask corner in the island structure.

5.1.3. Narrow Mask Effect on Bird's Beak Length

Fig.16 shows the bird's beak length versus nitride mask width in the hole structure. The mask width W_x in the x-direction is a variable. The mask width W_y in the y-direction is fixed. The experimental data for the bird's beak length L_y and L_x were obtained by the top view SEM photograph. The bird's beak length L_x keeps almost constant over all width of the nitride mask. The calculated result is consistent with the experimental data. The bird's beak length L_y is dramatically increased when the mask width W_x becomes narrower than $2.0 \mu m$. We call this phenomenon the narrow mask effect on the bird's beak length. At the corner of the mask edge in the hole structure, the bird's beak length is much enhanced than that at the mask edge far from the corner. The narrower the mask width W_x becomes, the bird's beak length

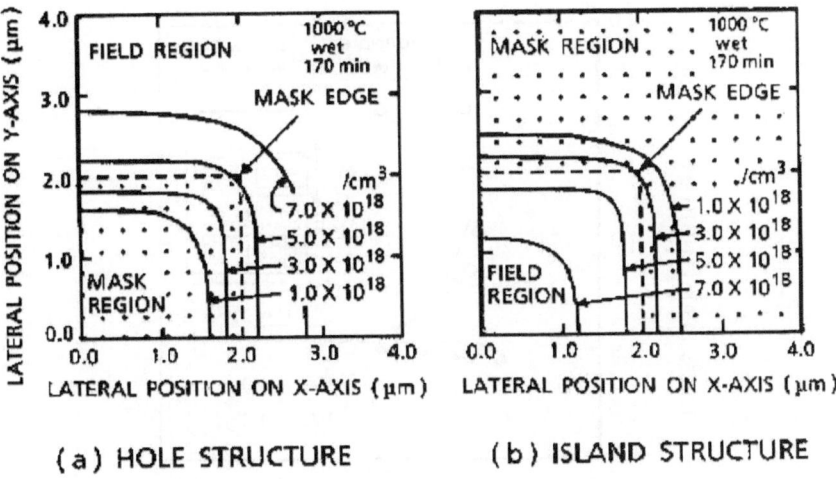

Figure 14: Oxidant distribution on silicon/oxide interface.

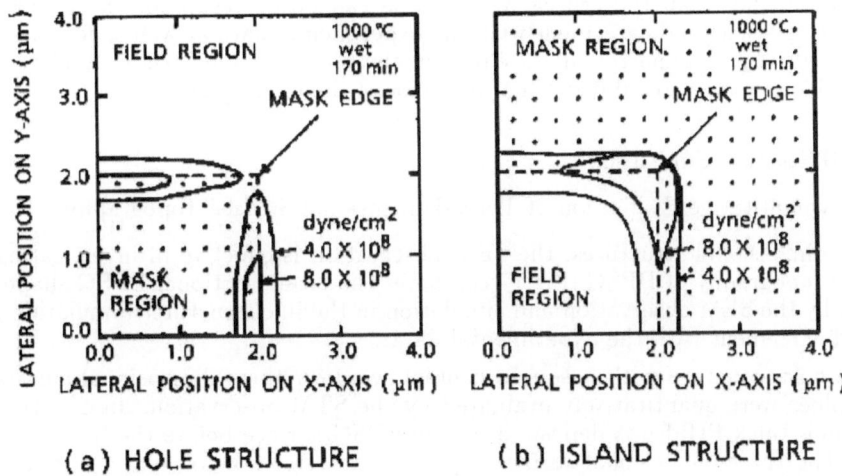

Figure 15: Pressure distribution on silicon/oxide interface.

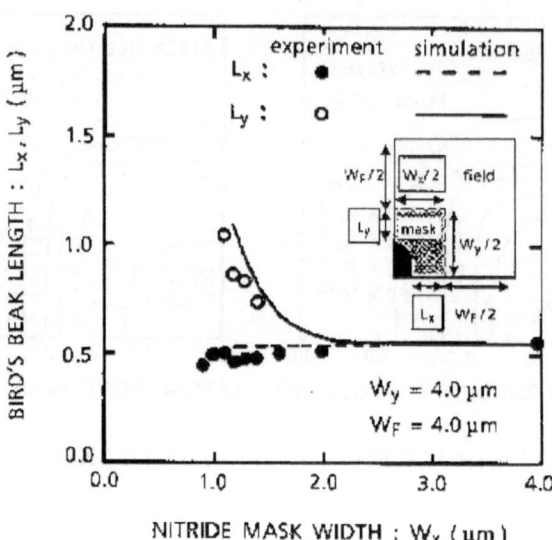

Figure 16: Narrow mask effect on bird's beak length.

is strongly affected by the lateral expansion of the oxide growth from the corner of the mask edge. As a result, the bird's beak length L_y is greatly increased when the distance $W_x/2$ between the center of the mask edge and the corner of the mask edge becomes smaller than the critical value. The calculated result for the bird's beak length L_y gives a good agreement with the experimental data as well as for the bird's beak length L_x. The narrow mask effect on the bird's beak length is simulated well by considering the three-dimensional behavior of the oxide growth

5.2. BPSG Flow Simulation

5.2.1. Quantitative Evaluation of Three-Dimensional Surface Topography

In two-dimensional structures, the SEM observation is effective in investigating the surface topography of BPSG. Fig.17 compares the cross- sectional BPSG shapes obtained by the SEM observation and simulation in the line structure. Simulation gives a good agreement with the experimental data.

In the hole structure with a square contact window, three-dimensional surface topographies were quantitatively evaluated by the STM observation. In experiments, the 5-nm- thick PtPd was deposited on the BPSG surface before the STM observation. The three-dimensional surface topographies after flow of 30min at 850°Cand 900°C were obtained by the STM observation as shown in Fig.18. In this case, the step height is 500 nm and the width of the square window W is $1.5\mu m$. The thickness of the deposited BPSG layer is 400 nm. The total impurity concentration is 17.5 mol%. The anneal is carried out for 30 minutes in a nitrogen ambient. The point A is located at the center of the contact window.

The STM observation allows a direct comparison of the BPSG surface profile with the simulation results. Fig.19 shows a direct comparison of the surface profiles obtained

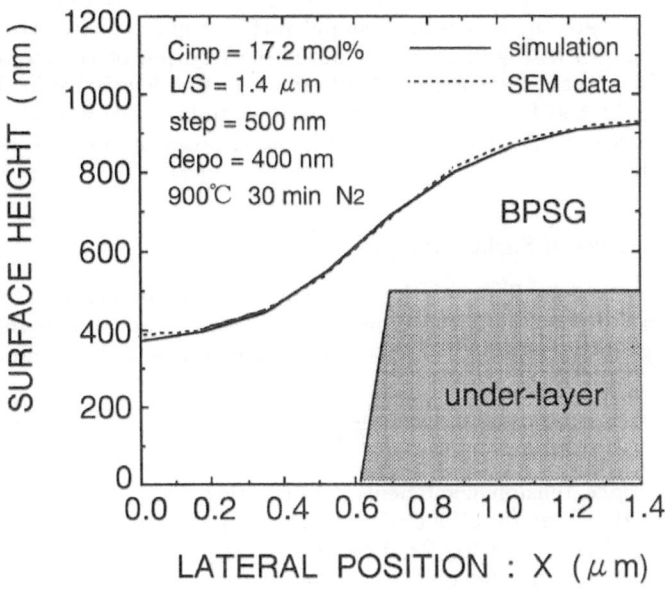

Figure 17: Direct comparison of BPSG surface profile in the line structure.

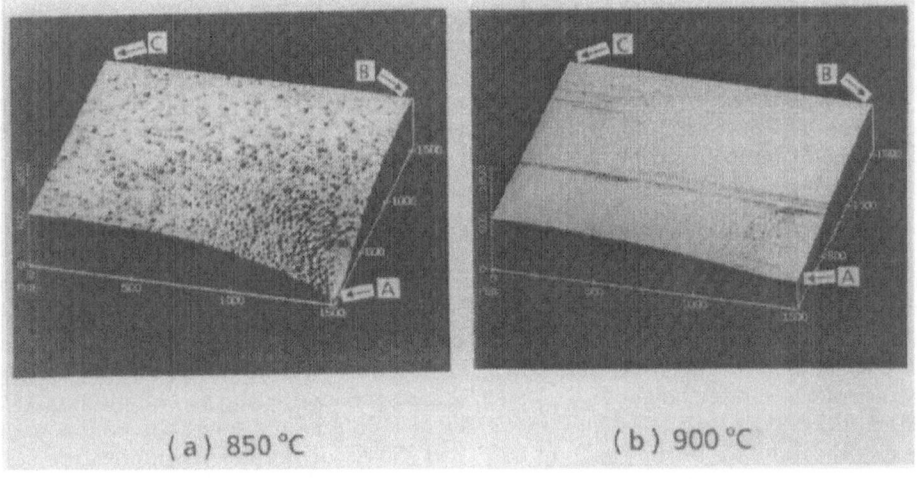

Figure 18: Three-dimensional surface topographies obtained by the STM observation.

by the STM in both the AB and AC direction at each temperature. The results clearly indicate that the surface height H_b at point B is larger than the surface height H_c at point C at 850°C and smaller than H_c at 900°C. The model simulates well this phenomenon for the three-dimensional surface topography of BPSG at each temperature. As shown in Fig.20, the calculated distribution of vertical component of the flow velocity reveals the significant difference of the flow behavior between at 850°C and 900°C. At 850°C, the surface at point A sinks down and the surface at points B and C raises up. At 900°C, the surface at A raises up and the surface at B and C sinks down.

5.2.2. Universal Curves of Surface Height

We found the characteristic flow length λ by using the present viscosity model. It is available for characterizing the BPSG flow behavior which depends on the temperature and impurity concentration. The characteristic flow length λ is defined as

$$\lambda = \frac{\gamma\, t}{\mu(T, C_{imp})} \tag{39}$$

where γ is the surface tension coefficient, t is the flow time and μ is the present viscosity model of BPSG as functions of the flow temperature T and total impurity concentration C_{imp}. Fig.21 and Fig.22 show the surface height versus the characteristic flow length λ in the line and hole structures, respectively. In the line structure as shown in Fig.21, it is confirmed that the experimental data in the wide range of the temperature and impurity concentration falls on the calculated universal curves of the surface height, which depend just on the initial structure of BPSG. The BPSG flow shows the different behavior of the surface height between small and large λ. When λ is smaller than about 2.0μm, the difference of the surface height H_{max} and H_{min} is increased. When λ is more than about 2.0μm, it turns to be decreased and the planarization is progressed.

For the hole structure as shown in Fig.22, we can further find the critical value λ_C, where the maximum surface position changes from B to C. This is because the surface at C continues to raise up while the surface at B sinks down. In the case of 850°C as shown in Fig.19(a), λ is smaller than λ_C. In the case of 900°C as shown in Fig.19(b), λ is bigger than λ_C. It is found that the BPSG flow behavior under various flow temperatures and impurity concentrations in a nitrogen ambient can be predicted and characterized by the universal curve using the present viscosity model.

5.3. Pattern Width Dependence of BPSG Flow behavior

To investigate the BPSG flow behavior in the various size structure, the simulation results are compared with the experimental data in the wide variety of size line and hole structure. Fig.23 shows the line and space dependence of the maximum and minimum surface height H_{max} and H_{min} in the line structure, respectively. The difference of the surface height H_{max} and H_{min} is decreased with the reduction of the line and space. The calculated surface height of BPSG is in a good agreement with the experimental data in the range of 850°C to 950°C for the flow temperature.

In the hole structure, the window width dependence of the maximum difference of the surface height is shown in Fig.24. At 900°C and 950°C, simulation results are in a good agreement with the STM data in the hole structure with the various size. At 850°C, there is a discrepancy between simulation and STM data, in particular,

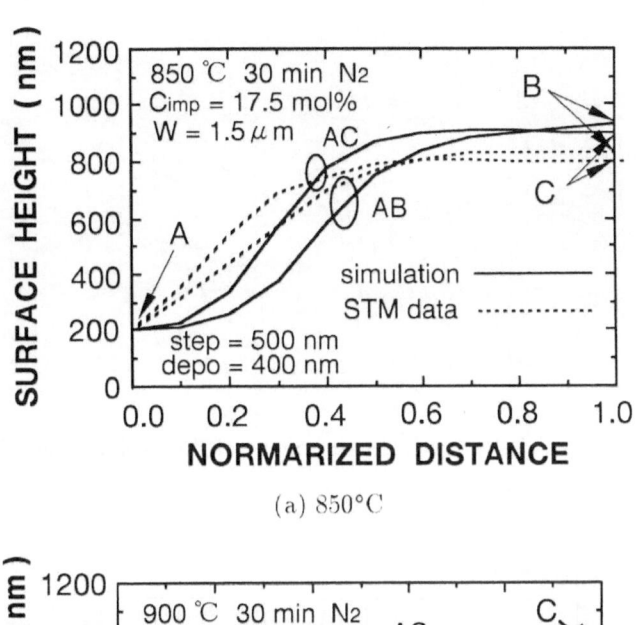

(a) 850°C

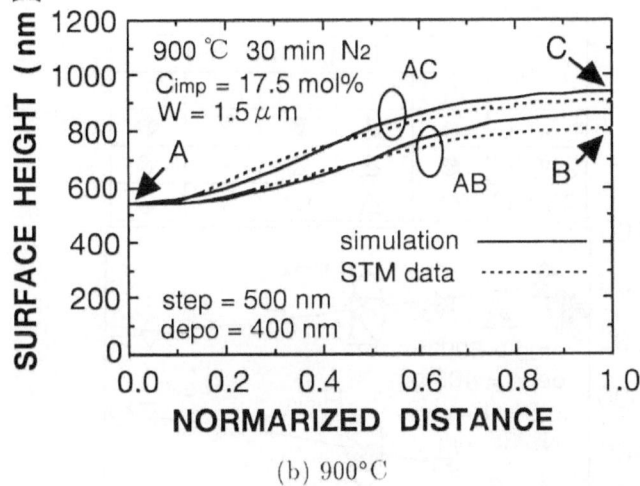

(b) 900°C

Figure 19: Direct comparison of the surface profile in the AB and AC direction in the hole structure.

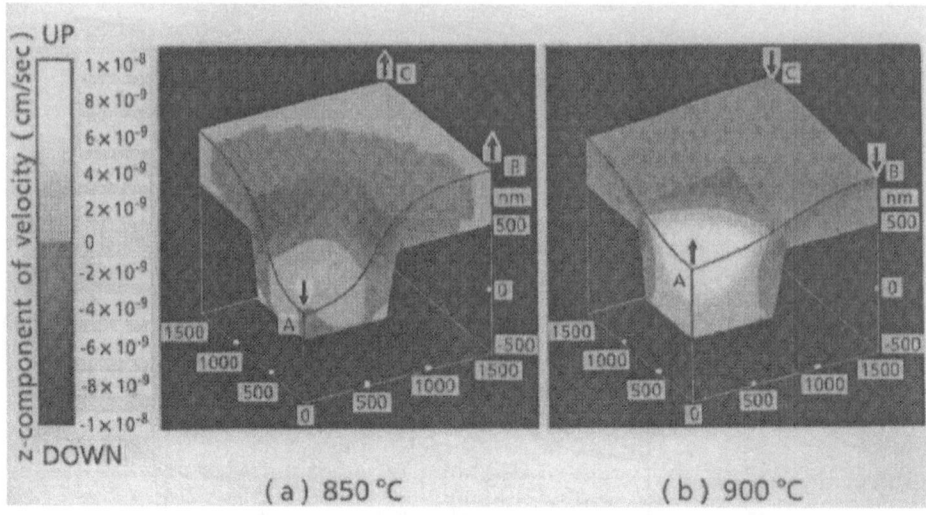

Figure 20: Distribution of the vertical component of flow velocity obtained by simulation.

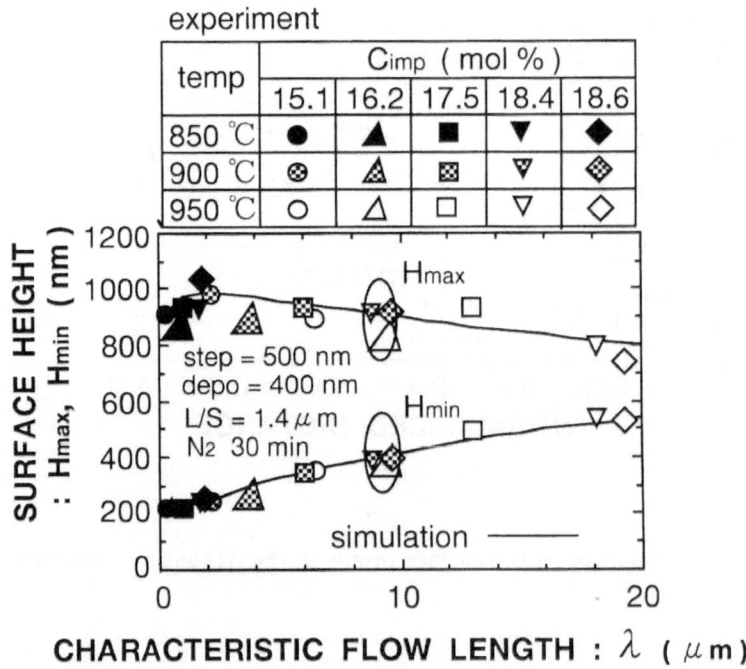

Figure 21: Universal curve of the surface height in the line structure.

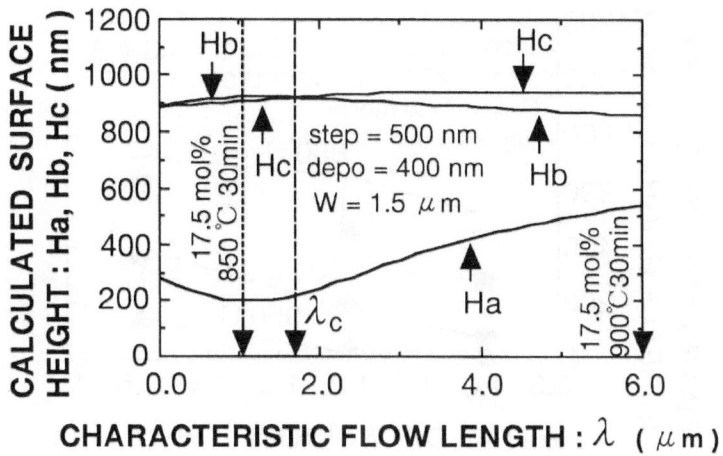

Figure 22: Universal curve of the surface height in the hole structure.

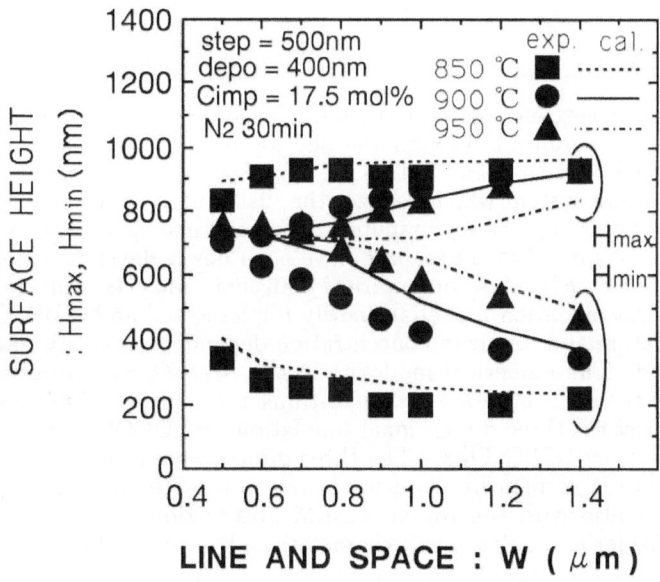

Figure 23: Surface hight versus line and space in the line structure

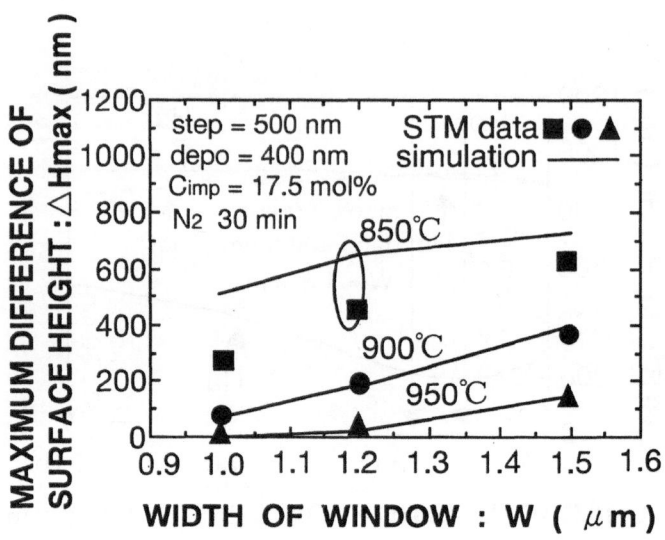

Figure 24: Surface hight versus window width in the hole structure

for the small window. This is because the maximum difference of the surface height at the low temperature is strongly affected by the initial deposition shape of BPSG layer.

6. Conclusion

We have described an overview of a three dimensional process simulator, SMART-P, which has been advanced through the development of the DRAM cells, CMOS processes and CCD devices. The SMART TCAD system has been reconstructed on the UNIX operating system which realizes the distributed environment using workstations and a vector type supercomputer. The task manager, a three-dimensional color graphics tool and a SIMS database have been newly developed and help to use the SMART-P effectively. A semi-empirical transient enhanced diffusion model and a stress-dependent oxidation model are newly implemented and a BPSG flow model using a temperature and impurity concentration dependent viscosity model has been newly developed. The numerical modeling in the Si/SiO_2 structure is successfully solved by using the efficient numerical algorithms and high-speed matrix solver. The SMART-P realizes the three-dimensional simulations of LOCOS with the stress effect and planarization by BPSG Flow. The three-dimensional calculated topography of LOCOS structure is verified by comparison not only with the cross-sectional SEM observation, but also with the top-view SEM observation. The three dimensional BPSG flow behavior is simulated and characterized by universal curves of the surface height. It is verified by the direct comparison of surface profile between simulated results and STM observation.

Acknowledgements

The authors wish to thank Dr. T. Takemoto, H. Esaki, Dr. M. Inoue, Y. Terui and Dr. K. Hatada for their encouragement. The authors also would like to acknowledge the contribution of many colleagues at the Semiconductor Research Center of Matsushita, especially A. Hiroki, K. Kurimoto, K. Misaka and K. Akamatsu for helpful discussion, I. Nakao, Y. Hirai, M. Nishio, K. Tateiwa, S. Imai and Y. Terai for their sample preparation and T. Kurumizawa for his STM measurements.

References

[1] S. Odanaka, H. Umimoto, M. Wakabayashi and H. Esaki, " SMART-P : Rigorous three-dimensional process simulator on a supercomputer, " *IEEE Trans. Computer-Aided Design*, vol. 7, no. 6, pp. 675-683, June 1988

[2] F. Fasching, S. Halama and S. Selberherr, *Technology CAD Systems*, Springer-Verlag Wien New York

[3] R. B. Marcus and T. T. Sheng, " The oxidation of shaped silicon surfaces, " *J. Electrochem. Soc.*, vol. 129, no. 6, pp. 1278-1282, June 1982

[4] K. Ohe, S. Odanaka, K. Moriyama, T. Hori and G. Fuse, " Narrow-width effects of shallow trench-isolated CMOS with n^+-polysilicon gate, " *IEEE Trans. Electron Devices*, vol. 36, no. 6, pp. 1110-1116, June 1989

[5] K. Kurimoto and S. Odanaka, " A T-gate overlapped LDD device with high circuit performance and high reliability, " *in Tech. Dig. of IEDM.*, pp. 541-543, 1991

[6] N. Shimizu, Y. Naito, Y. Itoh, Y. Shibata, K. Hashimoto, M. Nishio, A. Asai, K. Ohe, H. Umimoto and Y. Hirofuji, " A poly-buffer recessed LOCOS process for 256Mbit DRAM cells, " *in Tech. Dig. of IEDM.*, pp. 279-282, 1992

[7] H. Umimoto, S. Odanaka, I. Nakao and H. Esaki, " Numerical modeling of nonplanar oxidation coupled with stress effects, " *IEEE Trans. Computer-Aided Design*, vol. 8, no. 6, pp. 599-607, June 1989

[8] H. Umimoto, S. Odanaka and I. Nakao, " Numerical simulation of stress-dependent oxide growth at convex and concave corners of trench structures, " *IEEE Electron Device Letters*, vol. 10, no. 7, pp. 330-332, July 1989

[9] H. Umimoto, S. Odanaka and S. Imai, " A three-dimensional dynamic simulation of borophosphosilicate glass flow, " *in Tech. Dig. of Sympo. VLSI Tech.*, pp. 47-48, 1991

[10] H. Umimoto, S. Odanaka and S. Imai, " A 3-D BPSG flow simulation with temperature and impurity concentration dependent viscosity model, " *in Tech. Dig. of IEDM.*, pp. 709-712, 1991

[11] H. Umimoto and S. Odanaka, " Three-dimensional numerical simulation of local oxidation of silicon, " *IEEE Trans. Electron Devices*, vol. 38, no. 3, pp. 505-511, June 1991

[12] S. Odanaka, A. Hiroki, K. Ohe, K. Moriyama and H. Umimoto, " SMART-II: A three-dimensional CAD model for submicrometer MOSFET's, " *IEEE Trans. Computer-Aided Design*, vol.10, no.5, pp.619-628, May 1991

[13] H. Ryssel, *Ion Implantation Technique.*, New York: Springer-Verlag, 1982

[14] G. Fuse, H. Umimoto, S. Odanaka, M. Wakabayashi, M. Fukumoto and T. Ohzone, " Depth profiles of boron atoms with large tilt-angle implantation, " *J. Electrochem. Soc.*, vol.133, no.5, pp. 996-998, May 1986

[15] J. F. Gibbons, W. S. Johnson and S. W. Mylroie, *Projected Range Statistics, Semiconductor and Related Materials*, 2nd ed. Halsted Press, 1975

[16] D. Collard and K. Taniguchi, " IMPACT - A point-defect-based two dimensional process simulator: Modeling the lateral oxidation-enhanced diffusion of dopants in silicon, " *IEEE Trans. Electron Devices*, vol.ED-33, no.10, pp. 1454-1462, Oct. 1986

[17] R. B. Fair, J. J. Wortman and J. Liu, " Modeling rapid thermal diffusion of arsenic and boron in silicon, " *J. Electrochem. Soc.*, vol.131, no.10, pp. 2387-2394, Oct. 1984

[18] K. Taniguchi, D. A. Antoniadis and Y. Matsushita, " Kinetics of self-interstitials generated at the Si/SiO_2 interface, " *Appl. Phys. Lett.*, vol.42, no.11, pp. 961-963, June 1983

[19] S. Matsumoto, Y. Ishikawa and T. Niimi, " Oxidation enhanced and concentration dependent diffusions of dopants in silicon, " *J. Appl. Phys.*, vol.54, no.9, pp. 5049-5054, Sept. 1983

[20] D. Chin, S. Y. Oh, S. M. Hu, R. W. Dutton, and J. L. Moll, " Two-dimensional oxidation, " *IEEE Trans. Electron Devices*, vol. ED-30, no. 7, pp. 744-749, July 1983

[21] B. E. Deal and A. S. Grove, " General relationship for the thermal oxidation of silicon, " *J. Appl. Phys.*, vol. 36, no. 12, pp. 3770-3778, Dec. 1965

[22] J. A. Appels, E. Kooi, M. M. Paffen, J. J. H. Schatorji, and W. H. C. G. Veikuylen, " Local oxidation of silicon and its application in semiconductor device technology, " *Philips Res. Rep.*, vol. 25, pp. 118-132, 1970.

[23] D. B. Kao, J. P. McVittie, W. D. Nix, and K. C. Saraswat, " Two-dimensional thermal oxidation of silicon-II. Modeling stress effects in wet oxides, " *IEEE Trans. Electron Devices*, vol. ED-35, pp. 25-37, Jan. 1988

[24] W. Kern and G. L. Schnable, " Chemically vapor-deposited borophosphosilicate glasses for silicon device applications, " *RCA Review*, vol. 43, pp. 423-457, 1982

[25] P. Sutardja and W. G. Oldham, " Two-dimensional simulation of glass reflow and silicon oxidation, " *in Tech. Dig. of Sympo. VLSI Tech.*, pp. 39-40, 1986

[26] K. Nassau, R. A. Levy, and D. L. Chadwick, " Modified phosphosilicate glasses for VLSI applications, " *J. Electrochem. Soc.*, vol. 132, no. 2, pp. 409-415, 1985

[27] M. Yoshimaru and H. Matsuhashi, *Okidenki Kenkyu Kaihatsu*, vol. 53, no. 2, pp. 79-, 1986

[28] R. B. Penumalli, " A comprehensive two-dimensional VLSI process simulator program, BICEPS, " *IEEE Trans. Electron Devices*, vol. ED-30, no. 9, pp. 986-992, Sept. 1983

3-D Topography Simulation Using Surface Representation and Central Utilities

A. R. Neureuther, R.H. Wang, J.J. Helmsen, J.F. Sefler
E. W. Scheckler, R. Gunturi, and Rex Winterbottom

Dept. of Electrical Engineering and Computer Sciences,
University of California,
231 Cory Hall, Berkeley, CA 94720-1770

Abstract

There are major opportunities for new algorithms and system integration concepts in TCAD systems which can be met by developing centralized utilities. Suitable purpose-built high performance algorithms for surface representation based simulation developed in connection with SAMPLE-3D are described. An exploratory centralized services system called the Berkeley Topography Utilities has been developed for studying the continuum of flexible choices between reusing these purpose-built algorithms and robust general-purpose solid modeling operations. This system links code from SAMPLE-3D, SIMPL, the IBM Geometry Engine serving as a solid modeler, and a 2-D shock tracker. The BTU organization into hierarchial views, the use of surface direction monotonicity for speed enhancement, and a geometry tagging method for process trace-back are described.

1. Introduction

As process technology advances to smaller features sizes, larger chips and more limited depth of focus the interrelationships between parameters within a process step and between process steps are becoming very complex. Process simulation can help the technologist balance these trade-offs by providing understanding and quantitative assessment. While work will always be necessary to keep up with the physical mechanisms in processing, work is also needed on computational aspects such as those in representing and advancing the geometry. Simulation in 3-D faces tough problems in representing surfaces, in addressing sudden topological changes, in regularizing the resulting surfaces, and in advancing surfaces at fans or shock locations. Experts in both physical modeling and computational geometry can make contributions to TCAD. The key to taking advantage of this specialization and reusing the associated results is in establishing flexible organizational schemes which allow interfacing at primitive geometry through aggregate action levels. An interesting issue is the extent to which existing software and techniques from graphics and solid modeling can contribute. Purpose-built codes can also contribute and with modular approaches can be shared in various TCAD systems [1-19] and tools [20-36] as centralized utilities.

At Berkeley as part of SAMPLE-3D [37-43] we have developed software for representing 3-D device topography as triangulated surfaces which can be advanced and regularized. We have also developed an exploratory centralized services system called the Berkeley Topography Utilities [44-46] for studying organizational issues in combining purpose-built high performance algorithms with robust general-purpose solid modeler operations. This paper gives a perspective view of these capabilities. The general strategy of a modular utility approach arising out of application specific simulators is discussed in the following section. Then the capabilities of surface representation based simulation as currently available in SAMPLE-3D version 2.0 are presented. Alternative surface motion solution methods of cell, ray-trace and advection are compared and the key algorithmic concepts of surface representation based methods are discussed. A set of thin-triangle queries and procedures for clipping and regularizing triangulated surfaces including extraction of solids for linking to FastCap [47] and FastHenry [48] follows. The final section reports on the BTU organizational scheme which links the SAMPLE-3D capabilities together with those of the IBM Geometry Engine [49] acting as a solid modeler to carry out topography simulation.

2. Application Specific Simulators and Utilities for TCAD Systems

To survive in these changing times we at Berkeley have adopted a two pronged strategy of on one hand targeting world class utilities to solve CPU intensive issues in major TCAD systems and on the other hand sustaining our stand alone functionality to undertake collaborative application studies to assist technologists. While developing utilities does not seem as exciting as having visions about frameworks, utilities are in fact the stuff that frameworks are made of. The general approach is to work on break-through approaches to difficult computational or physical modeling aspects which are key to the application within our normal area of technology expertise. Of the emerging algorithms and codes the ones that make sufficiently strong contributions will be welcome and worthwhile to adapt to various TCAD frameworks. We are generally willing to write the wrappers in adapting the code and look forward to working with those systems which have sufficient flexibility to accept contributions to centralized utilities.

There are major opportunities for new concepts and algorithms to play a role as utilities in TCAD. Figure 1 depicts the physics of process and device on the left and integration level activities such as the TCAD, CIM and IC Design systems on the right. The utilities in the middle are envisioned as providing additional horsepower. For example, in topography simulation utilities for the difficult and specialized jobs of removing loops, regularizing surface panels and converting surfaces to solids could be developed. In impurity simulation utilities are needed for modifying the impurity representation to account for removal of regions in etching or addition of materials in deposition. The proof of concept circle with its connecting arrows is included to indicate that there are many new frontiers particularly involving interconnections between systems views such as in equipment characterization with CIM and layout and process flow with IC design.

Many of the utilities we are targeting at Berkeley emphasize the use of simulation based on representing surfaces by means of interconnected triangles of a nearly similar size. These and related concepts (such as monotonicity) can form a foundation for CPU efficient evaluation of surface self-shadowing and re-

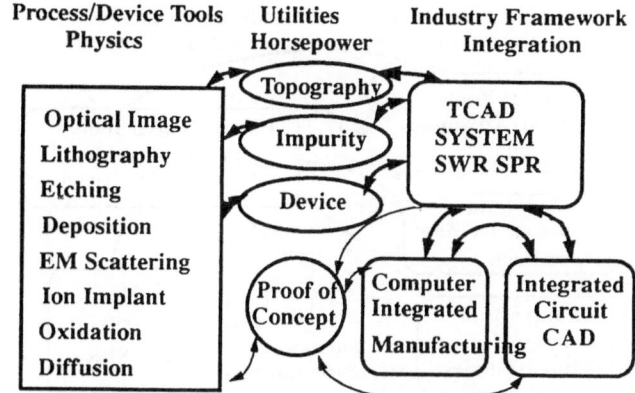

Figure 1: Opportunities for centralized topography, impurity, device
and system integration utilities in TCAD.

emission of material flux from the surface which are much more complex in 3-D. The use of a surface representation provides both the accuracy to describe the geometry and the computational efficiency in topography change of only analyzing the surface itself. This is particularly tricky at interfaces between materials and at locations of shocks and collisions. However, recent advances in deloop techniques (using an OctTree) and thin-triangle regularization which can cope with standing wave effects in photolithography demonstrate that robustness can be provided with little loss of efficiency. Effective adaptive surface representation based computational techniques in 3-D could thus be developed for coping with multiple material interfaces, rapid changes in shape or shadow, and CPU intensive problems in transport.

Based on the richness of these capabilities and our experience with the IBM Geometry Engine as a solid modeler we have embarked on the development of the Berkeley Topography Utilities (BTU) for use in TCAD systems. The general vision of this set of utilities is shown in Figure 2 . The center of the Figure is the surface representation which must be maintained as a valid surface. This representation can be acted on by the TCAD system itself, the user with their own tools, or the user through invoking the BTU views on the left side of the page. The user is given a flexible choice of the level at which to interact with the surface representation be it at a primitive geometry, an aggregate action or a call to auxiliary views of the geometry or process flow.

While the Berkeley Topography Utilities were developed to research organizational issues they are also intended to be useful in a variety of TCAD system environments. The initial goal was to include them in a common TCAD system based on the Semiconductor Wafer Representation (SWR) and Semiconductor Process Representation (SPR) [1]. Sematech championed the development of an SWR compliant TCAD Work Bench with an SWR server and several tool agents were written for exposure bleaching and development in lithography [2]. Figure 3 shows various components and contributions to this system. The proof of concept bubble and its connecting arrows indicates that there are many opportunities for academic exploration. However, the vision of the SWR and even the TCAD workbench architecture are apparently not viable as

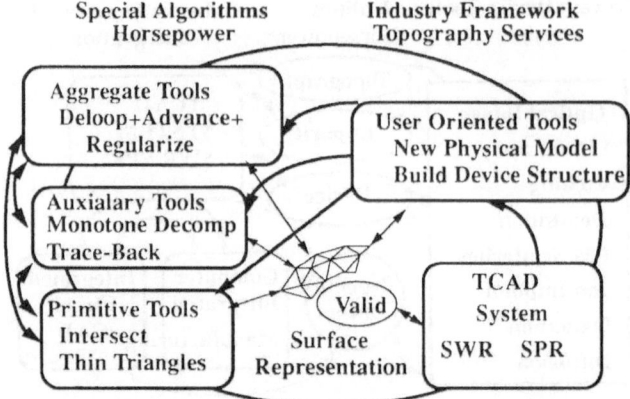

Figure 2: A flexible approach for allowing a surface representation to be checked for validity and operated on by a TCAD system, user oriented interface tools, and special high-performance utilities.

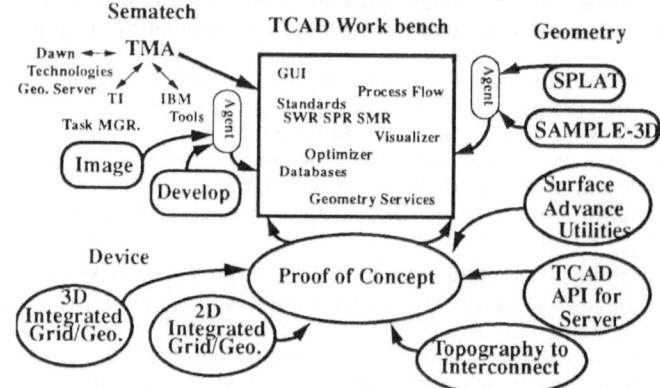

Figure 3: The Sematech TCAD Work Bench with contributed task manager and geometry server services, the integration of simulation tools through tool agents, and the opportunities for proof of concept development.

commercial products. As a result we have been asked to target our contributions at commercial versions of TCAD systems [3]. Currently there is also some hope of participating in a very general TCAD system through collaboration on joint SRC-National Labs simulation projects which will bring together many state-of-the-art mathematical and physical approaches. In addition there are also several inter-university collaborations which offer opportunities for contributing utilities.

3. SAMPLE-3D Overview

Many key capabilities for operating on surface based representations are available in the recent release of SAMPLE-3D version 2.0. The overall organization is shown by means of the IC process flow structure in Figure 4 . The lithography tool

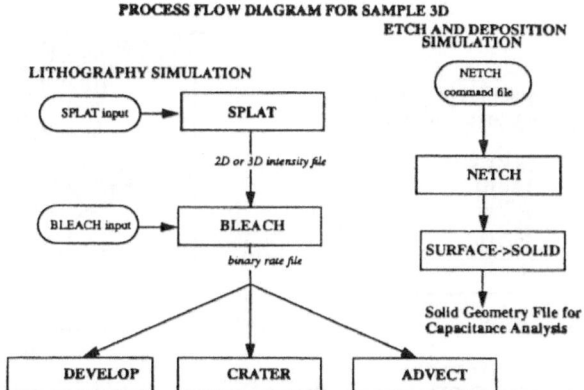

Figure 4: Process flow organizational view of SAMPLE-3D v2.0

set is on the left and consists of aerial image simulation with SPLAT[37] followed by the bleaching of the photoresist during exposure in BEACH[38]. Finally the dissolution can be simulated using surfaces (DEVELOP)[38,41,42], cells (CRATER)[39] or level-set methods (ADVECT) [42]. The right hand side shows surface based etching and deposition (NETCH) [40] and surface to solid extraction [43] for creating solid geometry descriptions which are compatible with FastCap [47] and FastHenry [48] for interconnect analysis.

An example of a SAMPLE-3D surface for photolithography is shown in Figure 5 .

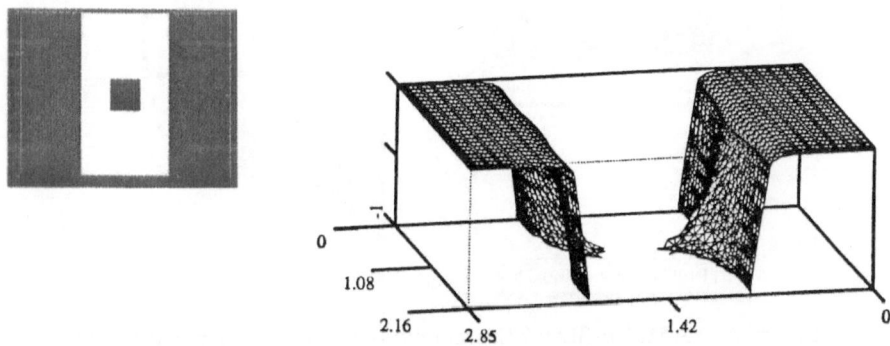

Figure 5: SAMPLE-3D v 2.0 projection printing resist profile in positive resist from a mask with an opaque defect.

Here an opaque defect between two features has reduced the intensity during exposure and created a linewidth variation. At the location of the defect advancing etch fronts from behind the defect and in front of the defect collide. SAMPLE-3D now has the capability to calculate the surface intersection, keep the valid surface, remove thin-triangles, and proceed with further etching.

Examples of surfaces in etching and deposition are shown in Figure 6 . SAMPLE-

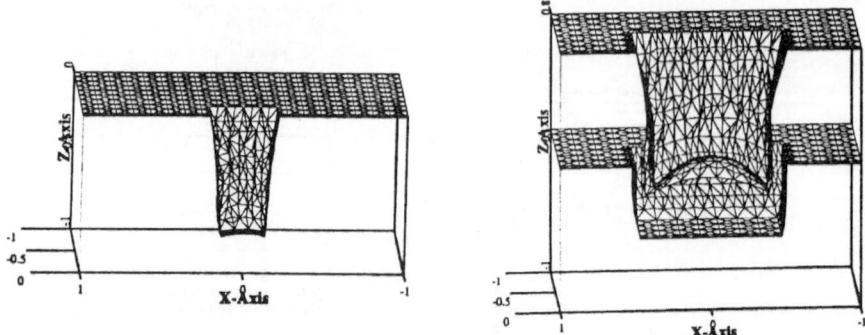

Figure 6:   SAMPLE-3D v 2.0 profiles from etching with ion reflection
and deposition.

3D allows physical effects such as self-shadowing, intra-surface radiosity, ion
reflection and surface migration, and crystal etching to be included. In addition to
the surface representation an auxiliary volumetric grid also carries inhomogeneous
attributes which can be accessed or calculated during processing. This auxiliary
data structure supports simulating silylated resist and material density variations
in deposition and etching. Figure 7 shows an example in which the density of

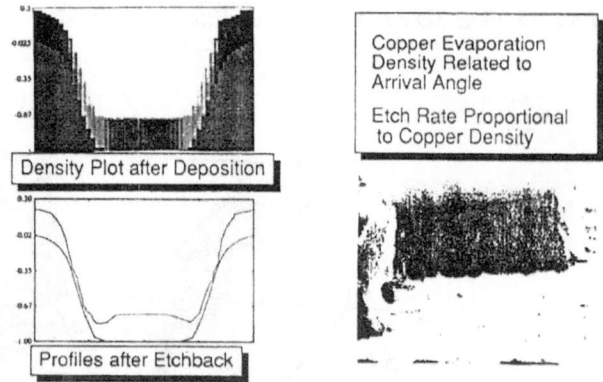

Figure 7:   SAMPLE-3D v 2.0 simulation of density in deposition and
the rapid removal of less dense material in etchback. SEM for Copper.

material depends on the surface orientation during deposition. It is lower in on the
sidewalls and is subsequently removed very rapidly in a short etch-back.

An example of solid extraction in SAMPLE-3D v 2.0 is shown in Figure 8. An
initial surface after lithography and plasma etching of a polysilicon layer is on the
left. This surface is a composite of the surfaces of the resist, polysilicon and oxide
which are exposed to the etchant. To extract the solids which represent the two

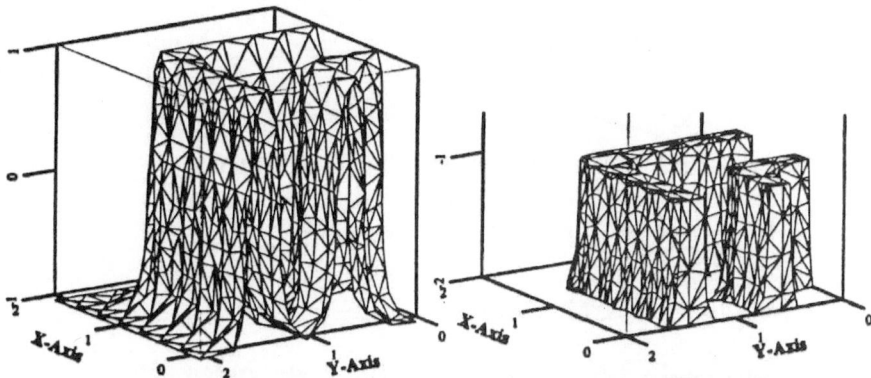

Figure 8: SAMPLE-3D v 2.0 extraction of solid polysilicon features
from an etched surface profile for linking to FastCap.

conducting pieces of polysilicon the initial surface must be clipped against a
surfaces representing the resist and oxide interfaces and against the sidewalls of
the bounding simulation domain. Each of these additional surfaces are first
triangulated, then clipped and the resulting small and thin triangles are then
removed. The resulting solids have 922 and 404 triangles which can be written to
a file for input to FastCap and FastHenry or other programs for analysis. The
mutual capacitance for these two physical features when the process bias and
profile shape are included is for this structure about 40% smaller than that
predicted by simply extruding the masks to form the features.

4. Dissolution Algorithms in SAMPLE-3D

We turn now to examine in greater detail the concepts and algorithms used in
SAMPLE-3D. The initial 3-D surface advancement by moving interconnected
nodes along etching rays was formulated as a vector equation by Paul Hagouel
[20], and developed into a predictor-corrector approach by Kenny Toh [37]. Ed
Scheckler extended this method to moving facets for etching and deposition [40].
He also added a cell development method called CRATER [39]. John Helmsen
addressed the speed of the surface self-intersection calculation and made the ray
tracing code of DEVELOP more robust by adding a spatial sorting based
(OctTree) deloop and thin-triangle regularization [41]. He also added the
ADVECT dissolution routine [42] which follows the method of Sethian [50,51] to
solve for level-set contours over a volume and thus avoids the complexity at shock
fronts with surface based methods.

The cell, surface and level-set advancement methods can be viewed as arising
from the Hamilton-Jacobi Equation as indicated in Figure 9 . Typical results for
developing a contact hole are shown in Figure 10 . Since far more time was spent
to develop the ray-trace compared to the cell and advection methods the accuracy,
CPU times and coding data reported here hardly represent a fair comparison. Yet
they do identify important issues. The major concern in the cell method is that of
getting sufficient accuracy without a slow down in estimating the surface normal
from the cell structure. The advection method trades simulation of a 2-D surface
moving with time to a function over a 3-D volume changing with time and thus

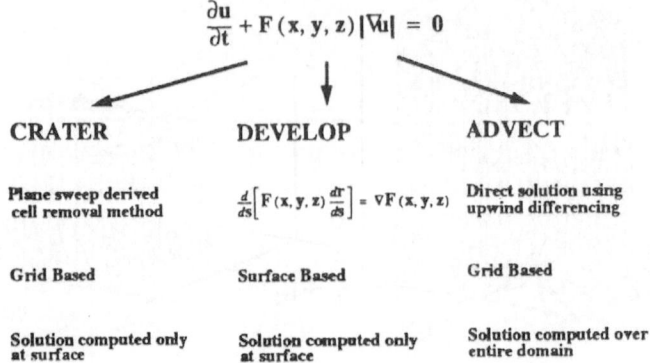

$$\frac{\partial u}{\partial t} + F\,(x, y, z)\,|\nabla u| = 0$$

CRATER	DEVELOP	ADVECT
Plane sweep derived cell removal method	$\frac{d}{ds}\left[F\,(x, y, z)\,\frac{dr}{ds} \right] = \nabla F\,(x, y, z)$	Direct solution using upwind differencing
Grid Based	Surface Based	Grid Based
Solution computed only at surface	Solution computed only at surface	Solution computed over entire domain

Figure 9:   Summary of the attributes of the cell (CRATER), surface (DEVELOP) and level-set (ADVECT) methods in SAMPLE-3D v 2.0.

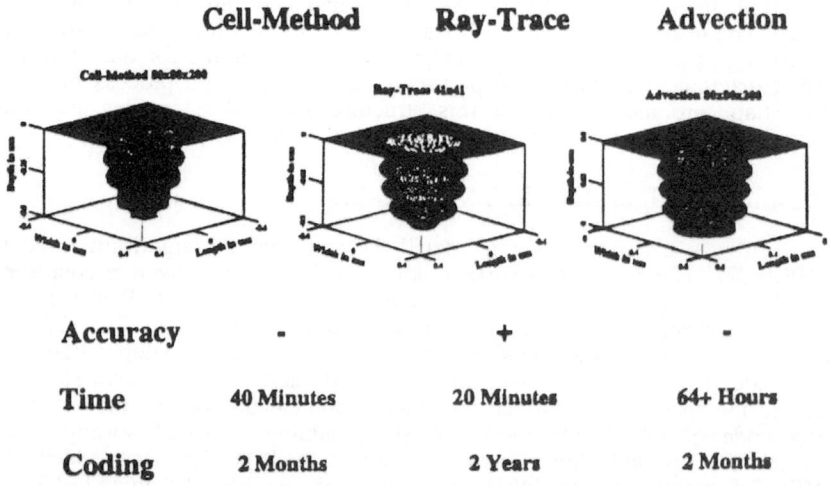

Cell-Method   **Ray-Trace**   **Advection**

	Cell-Method	Ray-Trace	Advection
Accuracy	-	+	-
Time	40 Minutes	20 Minutes	64+ Hours
Coding	2 Months	2 Years	2 Months

Figure 10:   SAMPLE-3D v 2.0 outputs for cell, surface and level-set methods and summary of accuracy, CPU time and coding effort.

suffers some additional CPU overhead. Advection can be sped up by cleverness in choosing the initial distribution of the unknown function over the 3-D domain and by concentrating the simulation accuracy only near that contour level of the function which represents the resist profile.

The problems in the surface representation methods are dealing with shock fronts at colliding surfaces, removing thin-triangles and sustaining surface flatness as rays traverse large distances over fields of discretized approximations of etch rates. The increase in local surface roughness or crenulation results from the formation of thin triangles which tend to become tilted with respect to the direction of the etch front advancement. An example of this effect at the bottom of a surface etched deep into a resist is shown in Figure 11 . The view on the left has

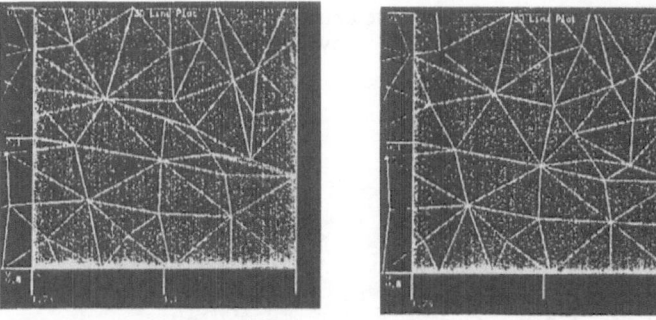

Advancement with Thin Triangles   **Advancement without Thin Triangles**

Figure 11:   Examples of thin triangles which evolve into crenulation
and their removal in SAMPLE-3D v 2.0.

several triangles with very low aspect ratios (height over base less than 0.25) located near the middle of the right hand side. By using routines to both remove thin triangles and adjust etch rates at local nodes to preserve more global measures of surface curvature this crenulation can be prevented. It is now possible to undertake photolithography problems with multiple shock fronts and large numbers of triangles as can be seen in Figure 1 2.

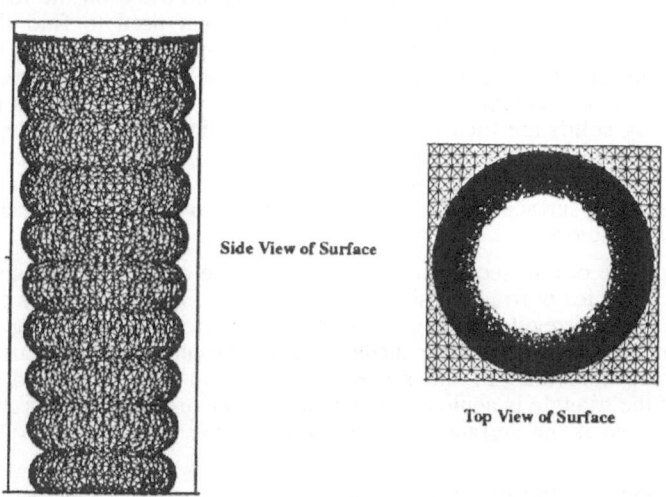

Side View of Surface

Top View of Surface

Figure 12:   SAMPLE-3D v 2.0 example of a high aspect ratio contact
hole with many standing wave fringes which have been delooped.

5. Extraction of Solids From Surfaces

The extraction of solids from surface representations as developed for SAMPLE-3D [43] is also based on an number of concepts and algorithms which are of interest as utilities in TCAD systems. The extraction requires intersecting arbitrary shaped triangulated surfaces. Figure 13 shows a special case of interest in which

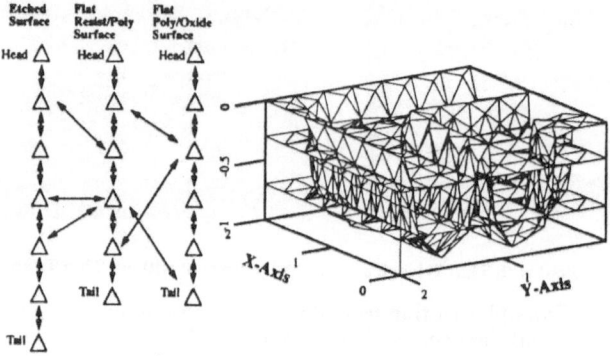

Figure 13:   Method of intersecting surfaces in SAMPLE-3D v 2.0
using link lists of triangulated, intersected, and refined surfaces and
cross-linking to assemble the extracted regions.

the intersecting surfaces are the planes of the resist/polysilicon and the polysilicon/oxide interfaces. These interface surfaces are first triangulated and each is put into a general linked list of triangles as indicated on the left side of Figure 13. The deloop algorithm of Helmsen is then used to find the exact intersection. Intersected triangles are divided into sub-polygons which is followed by triangulation of the sub-polygons. The three linked lists of surface triangles at this point also include the refined triangles. The correct surfaces forming the faces of the resulting solids are then grouped into a new link list. This is depicted in Figure 13 by the arrows showing the linkage among various portions of the triangles in each of the three initial surfaces. The clipping process is repeated using the bounding surfaces of the simulation domain to create the end caps on the polysilicon conductors.

As a consequence of intersecting the simulated surface with the clipping surfaces, problematic triangles of irregular size and shape arise. Figure 14 shows examples of the small and irregular triangles which result from the clipping process. Irregular long thin triangles appear shaded and small triangles are darkened. These triangles are problematic in the sense that they cause robustness difficulties especially if the triangle is acute to the extend that the area approaches zero. This is especially true if the surface were to continue advancement in a subsequent process step.

Several methods for re-tiling and optimizing triangulated surfaces are available and were considered. However, these methods are often based on minimizing an energy function dependent on distance between vertices and will either reposition the vertices of the original mesh or distribute a completely new set of vertices across the surface. For TCAD simulation nodes carry more than positional

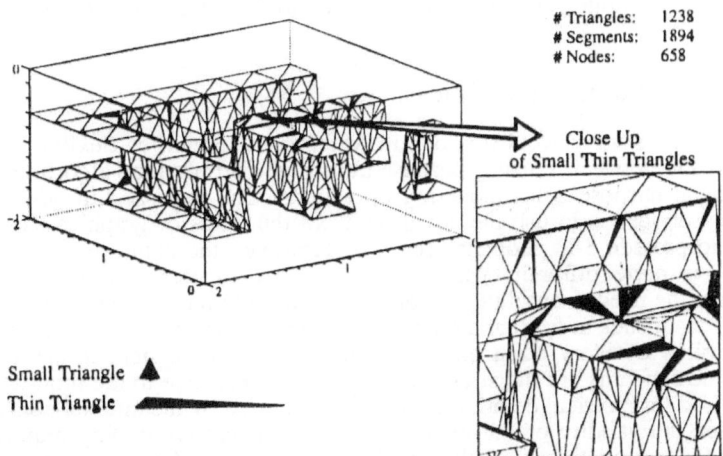

Figure 14: Examples of the small (dark) and thin (shaded) triangles which are generated in intersecting surfaces.

information. They also carry simulation fields such as etch rates, deposition rates, visibility attributes, direction vectors, and others. For this reason it is not desirable to significantly change the geometric position of nodes during a process step for the sake of thin triangle removal.

In order to solve the problem of removing all thin triangles throughout a surface without altering its major topographic features, a flexible query based algorithm was developed. This algorithm marches across the surface querying the individual triangles about their geometry and if their removal will alter the major features of the surface. Examples of the kinds or queries made on the geometry and the procedures which are enacted by heuristics are given in Figure 15 . Two key

Flexible Query Based Regularization

QUERRIES

SURFACE
Pinched?

TRIANGLE
Small?
Thin?
Ridge?
Edge?

SEGMENT
Ridge?
Edge?

NODE
Edge?
Corner?

REMOVAL QUERIES
Merge Alter?
DeLaunay Alter?

REMOVAL PROCEDURES
Merge Nodes
DeLaunay Flip
Disconnect

Figure 15: Examples of geometry queries and removal procedures used to remove thin triangles without altering the basic topography.

heuristics used in identification are the interior angle (thin if < 10 degrees) and if the triangle is a "ridge segment" which serves to define a major topographical feature.

6. Berkeley Topography Utilities

An ideal TCAD system would be one in which a continuum of flexible choices could be made between reusing purpose-built high-performance algorithms and robust general-purpose solid modeling operations. One approach to developing centralized services is to use the advances from the fields of graphics and solid modeling to provide 3-D geometry services. However, the nature of topography simulation is quite different which raises several concerns. In graphics the concerns are the need to emulate the role of hardware in removing hidden objects by plotting from back to front and the need to support the very massive calculation of the visibility of every point on the surface at every time step. In using solid modelers the concerns are the stress on the solid modeler due to the thousands of faces required for physical detail, the CPU time required by the internal constructs to guarantee the validity of the geometry, and the extensive coding required to map TCAD computation to solid modeler operations. Another approach in developing centralized services is to consolidate purpose-built 3-D geometry algorithms such as those in topography simulator for loop removal, surface regularization, and visibility into centralized utilities. This approach requires work towards rather than starting from, a common representation. It also attempts to install robustness retroactively, rather than implicitly inheriting it from constructs which underlay solid modeling.

The Berkeley Topography Utilities is an experimental TCAD system organization for 2D/3-D topography simulation. BTU was developed for researching the use of centralized services in TCAD systems. It is based on both purpose-built algorithms from SAMPLE-3D and the IBM Geometry Engine[49] as a solid modeler. Having both systems facilities performance evaluation and exploration of efficient access methods. While BTU was developed to facilitate research its interfaces and its purpose-built utilities will likely find applications in TCAD systems.

The BTU is a five layer object-oriented (C++) hybrid system [46] as illustrated in Figure 16 . The five layers are the Geometry Services at the bottom, the Primitive, Auxiliary, and Aggregate Views in the middle forming the BTU Hierarchial Views, and the top level which is the Application Development Views. The geometry services level needs both purpose-built servers for speed in surface advancement etc. and general purpose server backup using solids concepts. The Application Development Views enable rapid development of simulation programs by providing utilities for process flow management, simulation control, visualization, and physical model implementation. BTU Hierarchial Views allow flexible and efficient combinations of geometry services by organizing topography data into primitive, auxiliary, and aggregate views. Primitive utilities wrap geometry services, such as traversing cell-complex faces at the top surface, and perform data conversion, such as extracting surface mesh from a solid model. Auxiliary views contain utilities which improve geometry service access efficiency, such as monotone decomposition, or extend services with application specific semantics, such as tagging geometry with process and mask dependencies. Aggregate utilities, such as deformation and source visibility

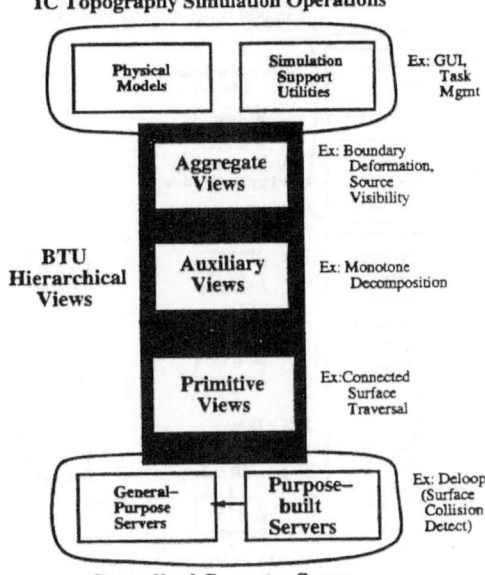

Figure 16: Organization of the Berkeley Topography Utilities into five
levels of views: Geometry Services, Primitive, Auxiliary, Aggregate
and Application Development.

utilities, perform geometrical computation needed in topography simulators by combining geometry services with numerical algorithms.

The Berkeley Hierarchial Views consists of about 41,000 lines of C++ code which integrate 36,000 lines of C code from SAMPLE-3D, 45,000 lines of C++ code from the IBM Geometry Engine, 36,000 lines of code from SIMPL System 5, and 2,400 lines of C code from a 2D shock tracker [52]. It is interesting to examine the code distribution with respect to the five layers of the organizational scheme. As shown in Figure 17 , almost two-thirds (and growing) of the SAMPLE-2D code involves purpose-built geometry algorithms to support ray-string surface advancement. This large amount of code at the primitive geometry level is an indication that there is a need for centralizing services. In the BTU system the Primitive Utilities layer is the largest (since that was the starting point) and the growth is taking place at the upper levels. The interface for each of the three geometry servers requires about 10,000 lines of code. In most cases the server wrapper development involves implementing straight forward data conversion algorithms, and code wrappers of similar servers are often reused. Most of the BTU development effort was concentrated on the 25% of the system software which contains aggregate and auxiliary utilities. Development of these utilities involved implementing sophisticated algorithms which map simulation computations to geometry services and partitioning topography data for efficient server access.

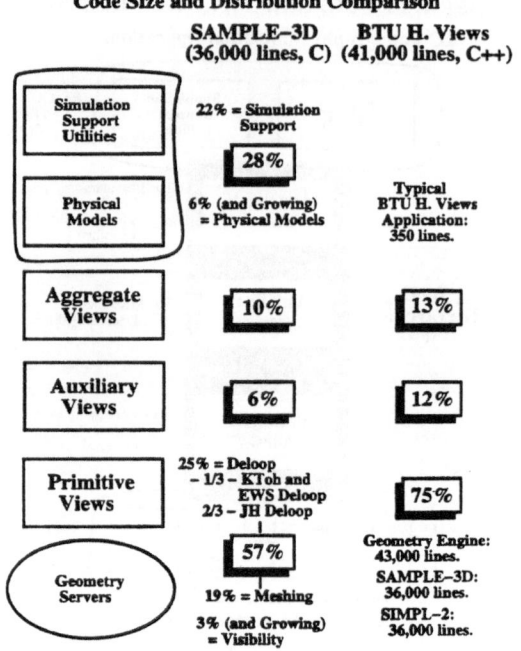

Figure 17:   Code size and distribution among the five view levels for
SAMPLE-3D and BTU.

The BTU system contains a set of tools for carrying out performance evaluation of geometry servers which has been very useful in assessing performance issues in adapting solid modelers to TCAD. For example, the IBM Geometry uses a modern winged-edge data structure with local connectivity and spatial decomposition for identifying intersections. It does not, however, allow localized geometry modification because it is difficult to guarantee correctness. As a result after every merge operation, the complete solid must be reconstructed at a considerable computational expense. This overhead can be seen in the simple Staircase test in Figure 18 . In the small grain approach each of N blocks is merged one by one and the time increases more than just linearly with N. In the large grain approach blocks are merged by sweeping out connected sets of size sqrt N. Here the time drops by the number of merges and over a factor of 37 is saved for 1,600 cubes. Thus using the robustness of solid modeling system can be thought of as coming at the cost of a few seconds per merge operation due to the fact that the modeler must check after every single operation to be sure the resulting solid is a valid geometry.

One technique for decreasing the number of merges and hence the CPU time in using a solid modeler is to increasing the granularity of merges through techniques such as monotone decomposition of the surface[45]. Figure 19 shows how with an auxiliary view a key hole trench might be subdivided in to regions of connected triangles with restricted orientation. If the surface orientation is sufficiently

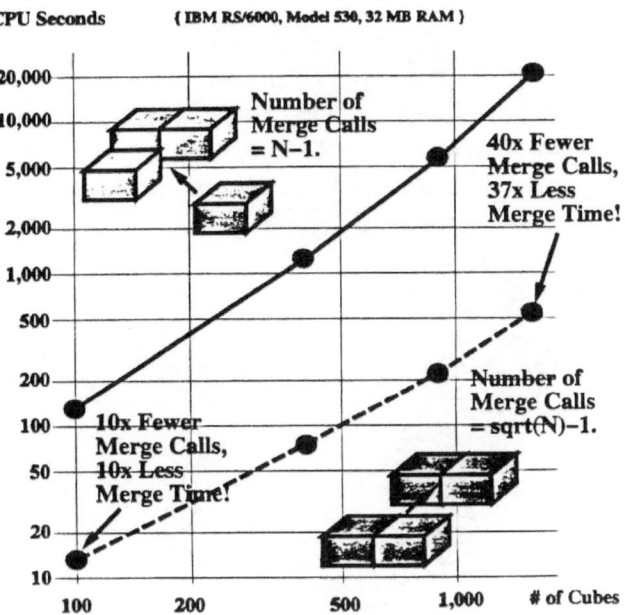

Figure 18: Results of a staircase merge test on the IBM Geometry
Engine which reveals that the internal validation of the geometry after
each merge in a solid modeler makes the CPU time dominated by the
number of merges and not the complexity of the geometry.

restricted then each of the individual nodes on the surfaces can be advanced
without concern for shocks or intra-surface collisions. These surfaces might be
advanced with very simple algorithms and then merged into a valid solid with only
a handful of merges by a solid modeler. This auxiliary view of the topography
would allow a solid modeler to take as little as one minute to rigorously resolve
abrupt topological surface changes. This monotone surface auxiliary view along
with the spatial decomposition of an OctTree might also be extended to speed up
visibility calculations in purpose-built algorithms.

Another interesting construct in BTU is the auxiliary process flow data structure
which allows the user to select a face in the display and determine the subset of the
process flow which created the topographical feature. The data structure attaches
and propagates SIMPL process ID's onto evolving features after each process step
as shown in Figure 20 . At the polysilicon lithography and etch process the
polysilicon has pointers attached to the lithography, its associated mask and then
the RIE process. The pointers to the lithography and mask are shown as dotted but
still remain even though the resist has ben ashed. When oxide is deposited it is
flagged as being influenced by the polysilicon layer underneath it. When the metal
is deposited it is flagged as being influenced by the underlying oxide. When the
metal is etched it is flagged by the metal etching process. Note that for the view in
the current cross section there is no lithography and masking of the metal so these
flags do not appear. To determine which process steps influenced the shape of the

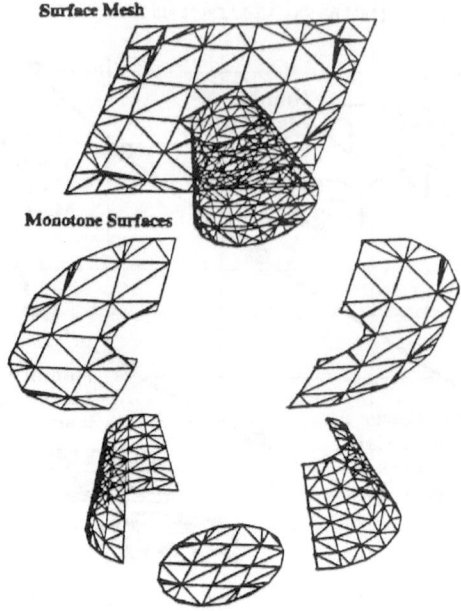

Figure 19: Example of using monotonicity of surface direction to divide a 3-D key hole shaped trench into subsections which may be advanced by simple code and then merged by a solid modeler.

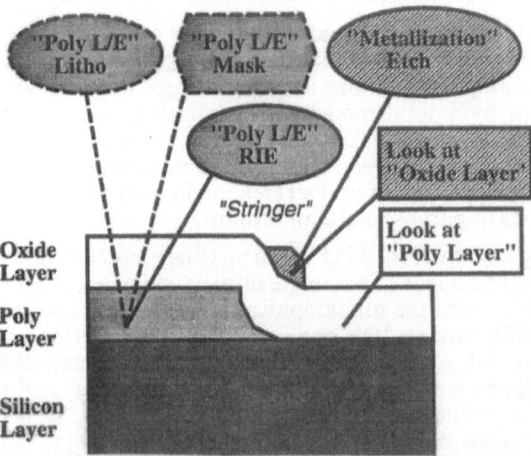

Figure 20: BTU topography propagation trace-back for determining all process steps which influenced the shape or properties of a region.

metal stringer the procedure is to follow the pointers backward. This is done by first looking at the masks and processes associated with the metal layer, then following the pointer to oxide and looking at the masks and processes which influenced the oxide layer, and finally going to the polysilicon layer and determining the masks and processes which influenced its shape. In this manner all possible influences can be diagnosed and some filtering of extraneous influences can be made by choosing the window small enough to isolate the particular topography under consideration.

7. Summary

There are major opportunities for new algorithms and system integration concepts in TCAD systems which can be met by developing centralized utilities. For 3-D topography simulation utilities based on surface representation for advancement, deloop, regularization and solid extraction are available. Using an exploratory centralized services system called the Berkeley Topography Utilities we are studying the continuum of flexible choices between reusing purpose-built high-performance algorithms and robust general-purpose solid modeling operations.

Developments in surfaces representation simulation of photolithography can now advance surfaces of over 10,000 triangles and cope with self-intersections, thin triangles at intersections, and crenulation. In extracting solids from simulation surfaces the problem of removing all thin triangles throughout a surface without altering its major topographic features has been solved using a flexible query based algorithm. The bottleneck in using solid modelers is in its validation of the structure after each merge. This can be overcome by sorting the 3-D structure into a small number of sets of faces with monotone orientation which can individually be advanced into solids without shocks or collisions and then using the robustness of the solid modeler to merge the collection of solids. An auxiliary geometry tagging method was introduced for allowing a user to select a face in the display and determine the subset of steps in the process flow which created or altered that topographical feature.

Acknowledgment

The research highlighted here was supported in part by the Semiconductor Research Corporation grant 94-MC-500, ARPA Advanced Lithography Program, ARPA iMEMS project, and Technology Modeling Associates and the California State Micro Program, for integration of Berkeley tools in the TCAD Work Bench.

References

[1] D. Boning et al., "Developing and Integrating TCAD Applications with the Semiconductor Wafer Representation," NUPAD IV, IEEE92TH0424-2, May 1992.

[2] For information on the TCAD Work Bench contact Sematech in Austin, TX.

[3] TCAD tools are available from Technology Modeling Associates, Silvaco, Dawn Technoloies, SIGMA-C, FINLE Technologies, Integrated Systems Engineering, Vector Technologies, and others.

[4] W. G. Oldham, S. N. Nandgaonkar, A. R. Neureuther, and M. M. O'Toole, "A General Simulator for VLSI Lithography and Etching Processes: Part I - Application to Projection Lithography," IEEE Trans. on Electron Devices, Vol. ED-26, No. 4, pp. 717-722 April 1979.

[5]   K. Lee, Y. Sakai and A.R. Neureuther, "Topography Dependent Step Coverage Resistance Simulation for VLSI Design," 1982 Symposium on VLSI Technology, Proceedings pp. 61-62, Oiso, Japan, Sept. 1-3, 1982.

[6]   G.M. Koppelman et al., IBM J. Res. Develop. Vol. 27, pp. 149-163, Mar. 1983.

[7]   C.P. Ho et al., IEEE Trans. on Electron Devices, Vol. ED-30, No. 11, pp. 1438-1453, Nov. 1983.

[8]   J. Lorenz et al., IEEE Trans. on Electron Devices, Vol. ED-32, No. 10, pp. 1977-1986, Oct. 1985.

[9]   C.H. Corbex, A.F. Gerodolle, S.P. Martin, and A.R. Poncet, IEEE Trans. CAD. vol. 7 no.4, pp. 489-500, April 1988.

[10] S.G. Duvall, IEEE Trans. on Computer-Aided Design, Vol. CAD-7, No. 7, pp., 41-754, Jul. 1988.

[11] K. Kato, et al. "A Supervised Simulation System for Process and Device Designs based on Geometrical Data Interface," IEEE Transactions on Electron Devices, vol ED-34, pp. 2049-2058, Oct. 1987.

[12] Alexander S. Wong, "An Integrated Graphical Environment for Operating IC Process Simulators," M.S. Thesis, University of California, Berkeley, May 1989. UCB/ERL M89/67.

[13] D.C. Cole, et al., Solid-State Electronics, vol. 33, no.6, pp. 591-623, 1990.

[14] P. Lamb, C. Hegarty, N. Hitschfeld, W. Fichtner, "Generating Solid Models for VLSI Process and Device Simulation," NUPAD-IV, Seattle, WA, May 1992, pp. 175-180.

[15] M.R. Pinto, D.M. Boulin, C.S. Rafferty, R.K. Smith, W.M. Coughran,Jr., I.C. Kizilyalli, and M.J. Thoma, "Three-dimensional characterization of bipolar transistors in a submicron BiCMOS technology using integrated process and device simulation," IEDM'92 Tech Digest, Dec 1992, pp 923-926.

[16] F. Fasching, W. Tuppa, S. Selberherr, IEEE Trans. CAD. vol. 13, pp. 72-81, Jan 1994.

[17] Z.H. Sahul, R.W. Dutton, and M. Noell, "Grid and Geometry Techniques for Multi-Layer Process Simulation," Proc. of SISDEP-5, Vienna, Austria, Sept. 1993, pp 417-420.

[18] C. Yang, M.D. Giles, "Architecture and Implementation of 3D Field Support for Semiconductor Wafer Representation," Proc of NUPAD V, June 1994.

[19] M. Law, World Wide Web Document URL: http://www.eel.ufl.edu/~law/manual/Intro.html.

[20] P.I. Hagouel and A.R. Neureuther, "Modeling of X-ray Resists for High Resolution Lithography", ACS Organic Coating and Plastics Preprints, Vol.35, No.2, pp.258-265, August 1975 (for 3-D vector ray-trace formulation) and in Ph.D. Thesis, UC Berkeley 1976.

[21] F. Jones and J. Paraszczak, "RD3D (Computer Simulation of Resist Development in Three Dimensions)," IEEE Trans. Electron Devices, vol. ED-28, no. 12, pp.1544-1552, December 1981.

[22] T. Matsuzawa, T. Ito, and H. Sunami, "Three-dimensional Photoresist Image Simulation on Flat Surfaces," IEEE Trans. Electron Devices, vol. ED-32, no. 9, pp. 1781-1783, September 1985.

[23] L. Jia, W. Jian-kun and W. Shao-jun, "Three-Dimensional Development of Electron Beam Exposed Resist Patterns Simulated Using Ray Tracing Model," Microelectronics Engineering, vol. 6, pp. 147-151, 1987.

[24] A. Moniwa, T. Matsuzawa, T. Ito, and H Sunami, "A Three-Dimensional Photoresist imaging process Simulator for Strong Standing-Wave Effect Environment," IEEE Trans on CAD, vol. CAD-6, no. 3, pp. 431-437, May 1987.

[25] Y. Hirai, S. Tomida, K. Ikeda, M. Sasago, M. Endo, S. Hayama, and N. Nomura, "Three Dimensional Process Simulation for Photo and Electron Beam Lithography and Estimation of Proximity Effects," Symposium on VLSI Technology, Digest of Technical papers, p. 15, 1987.

[26] E. Barouch, B. Bradie, H. Fowler and S. Babu, "Three-Dimensional Modeling of Optical lithography for Positive Photoresists," Interface '89: Proceedings of KTI Microelectronics Seminar, pp. 123-136, Nov. 1989.

[27] J. Bauer, W.Mehr, and U. Glaubitz, "Simulation and Experimental Results in 0.6 um Lithography using an i-Line Stepper," Proceedings of SPIE: Optical/laser Microlithography III, vol. 1264, pp. 431-445, march 1990.

[28] W. Henke, D. Mewes, M. Weiss, G. Czech, and R. Schiessl-Hoyler, "Simulation of Defects in 3-Dimensional Resist profiles in Optical Lithography," Microelectronics Engineering, vol. 13, pp. 497-501, 1991.

[29] K.K.H. Toh, A.R. Neureuther and E.W. Scheckler, "Three-Dimensional Simulation of Optical Lithography," SPIE Vol. 1463, pp. 356-367, March, 1991.

[30] K. Lee, Y. Kim and C. Hwang, "New Three-Dimensional Modeling of Optical Lithography for Positive Photoresists," 1991 International Workshop on VLSI Process and Device Modeling, pp. 44-45, Oiso, Japan. may 26-27, 1991.

[31] M. Komatsu, "Three Dimensional Resist profile Simulation," SPIE optical/Laser Microlithography VI, vol. 1927, pp. 413-426, 1993.

[32] M. Fujinaga, N. Kotani, M. Shirahata, H. Genjo, T. Katayama, T. Ogawa, Y. Akasaka, "Three Dimensional Topography Simulation Model Using Diffusion Equation," IEDM Technical Digest, pp. 332-335, Dec. 1988.

[33] J. McVittie, J. Rey, L-.Y. Cheng, M.M. IslamRaja, and K.C. Saraswat, "LPCVD Profile Simulation Using a Re-Emission Model," IEDM Technical Digest, pp. 917-920, December 1990.

[34] J. Pelka, "Simulation of Ion-Enhanced Dry-Etch Processes," Microelectronics Engineering, vol. 13, pp. 487-491, 1991.

[35] T.S. Cale, T.H. Gandy, M.K. Jain, M. Ramaswami, and G.B. Raupp, "A General Model for PVD Deposition," Proceedings Eighth International VLSI Multilevel interconnect Conference, pp. 350-352, Santa Clara, June 11-12, 1991.

[36] T. Smy, R.N. Tait, K.L. Westra, M.J. Brett "Simulation of Density Variation and Step Coverage for Via Metallization," Proceedings IEEE V-MIC, "Santa Clara, CA, pp. 292, June 1989.

[37] K.H. Toh and A.R. Neureuther, "Identifying and Monitoring Effects of Lens Aberrations in Projection Printing," SPIE Proceedings, Vol. 772, pp. 202-209, 1987.

[38] K.K.H. Toh, A.R. Neureuther and E.W.Scheckler, "Algorithms for Simulation of Three-Dimensional Etching," IEEE Trans. CAD, Vol. CAD-13, No 5, pp. 616-624, May 1994.

[39] E.W. Scheckler, N.N. Tam, A.K. Pfau and A.R. Neureuther, "An Efficient Volume Removal Algorithm for Three-Dimensional Lithography Simulation with Experimental Verification," IEEE Trans. CAD, Vol. CAD-12, No. 9, pp. 1345-1356, Sept. 1993.

[40] E.W. Scheckler and A.R. Neureuther "Models and Algorithms for Three-Dimensional Simulation with SAMPLE-3D," IEEE Trans. CAD, Vol. CAD-13, No. 2, pp. 219-230, Feb 1994.

[41] J.J. Helmsen, E.W. Scheckler, A.R. Neureuther and C.H. Sequin "An Efficient Loop Detection and Removal Algorithm for 3-D Surface-Based Lithography Simulation," NUPAD-IV Technical Digest, pp. 3-8, 1992.

[42] J.J. Helmsen, "A Comparison of Three Dimensional Photolithography Simulators," Ph.D. Thesis, University of California, Berkeley, December 1994, and ERL memorandum No. UCB/ERL M95/25 April, 1995.

[43] John H. Sefler and Andrew R. Neureuther, "Extracting Solid Conductors from a Single Triangulated Surface Representation for Interconnect Analysis," Submitted to IEEE Trans. Semiconductor Manufacturing, 1995.

[44] R.H. Wang, A. Gabara, A.R. Neureuther, "BTU - Berkeley Topography Utilities for Linking Topography and Impurity Profile Simulations," NUPAD-IV, Seattle, WA, pp. 225-230, May 1992.

[45] Wang, M.S. Karasick, and A.R. Neureuther, "Computational Evaluation of Three-Dimensional Topography Process Simulation Components," International Workshop on VLSI Process and Device Modeling (VPAD), Kyoto, Japan, May 1993, pp. 95-96.

[46] Wang, and A.R. Neureuther, "Efficient and Innovative Use of Three-Dimensional Geometry Services in IC Topography Simulation," International Symposium on VLSI Technology, Systems, and Applications (VLSI-TSA), Taipei, Taiwan, ROC, June 1995.

[47] K. Nabors, and J. White, "FastCap: A Multipole Accelerated 3-D Capacitance Extraction Program," IEEE Trans. on Computer Aided Design, vol. 10, no. 11, pp. 1447-1459, Nov. 1991.

[48] M. Kamon, M.J. Tsuj, and J. White, "FastHenry, A multipole-Accelerated 3-D Inductance Extraction program," Proceedings of the ACM/IEEE Design Automation Conference, Dallas, June 1993.

[49] M. Karasick and D. Lieber, ACM Symp. on CAD and Foundations of Geom. Modeling, pp. 15-25, June 1991.

[50] J. Sethian, "Analysis of Flame Propagation," Ph.D. Dissertation, University of California at Berkeley, 1982.

[51] J. Sethian, "Numerical Algorithms for Propagating Interfaces: Hamilton-Jacobi Equations and Conservation Laws," Journal of Differential Geometry, pp. 131-1161, 1990.

[52] S. Hamaguchi, M. Dalvie, R.T. Farouki, and S. Sethuraman, "A Shock-Tracking Algorithm for Surface Evolution under Reactive-Ion Etching," IBM Research Report, RC18283

Three-Dimensional Simulation
of Thermal Processes

Mark E. Law and Stephen Cea

Dept. of Electrical Engineering
University of Florida
Gainesville, Fl. 32611-6200
law@tcad.ee.ufl.edu, smc@tcad.ee.ufl.edu

Abstract

Three-dimensional simulation of thermal processes is performed using the Florida Object Oriented Process Simulator (FLOOPS). Algorithms for three-dimensional grid update, moving boundaries, and solution of equations are described. The major challenge in building a 3D simulator is addressing the grid requirements, and the approach used in FLOOPS will be described. As process simulators move to three-dimensions, new parts of the simulator become CPU limiting. Some of these problems will be discussed and addressed. Since this is an area of on-going work, our current research directions will also be discussed.

Introduction

As device and isolation structures are scaled, there are new problems that are presented that are purely three-dimensional in nature. The corner of the mask has become an important consideration in technology development. Since we are incapable of probing these corners experimentally, it has become increasingly important to develop physical insight into the corners through modeling and simulation. Three-dimensional device simulators have been around for some time, but three-dimensional process simulators are only now being developed in earnest.

Developing a three-dimensional thermal simulator from a two-dimensional code is primarily a challenge in data representation and numerical analysis[1]. The physics of thermal problems is largely unchanged for the change in dimension. This is completely different from the change from one to two dimensions, where simplifications to the physics had been exploited in one dimension. Two-dimensional codes had to handle more complicated physics, which is still not completely modeled and understood[2, 3]. Fortunately, this is not the case for increasing to three dimensions.

Florida Object Oriented Process Simulator (FLOOPS) is an excellent platform for developing a three-dimensional simulator. It contains most of the physics necessary for two-dimensional modeling. Since the grid, refinement, and assembly algorithms are object-oriented, it is straight forward to extend them to three-dimensions. We will present the FLOOPS data representation first, followed by a description of the grid algorithms used for two- and three-dimensional simulation. As we developed the simulator, we have discovered several interesting computational issues that will be also be discussed. Finally, we present some three-dimensional simulation results.

Object Oriented Representation

The mesh and grid representation are object oriented[4, 5], which provides a distinct advantage in developing a multi-dimensional process simulator. Figure 1 shows the mesh representation objects. The FieldServer object contains all the grid and spatial information. The FieldServer contains a list of Coordinate objects and Mesh objects. The Coordinate object represents a single position in space. The Coordinate object includes pointers to one or more Node objects. Each Mesh object incident on a coordinate will have a different node, so that boundary Coordinates point to multiple Node objects and internal Coordinates point to only one Node object. This approach is the same as used in SUPREM-IV[6], and allows for abrupt changes in the solution values when the material changes.

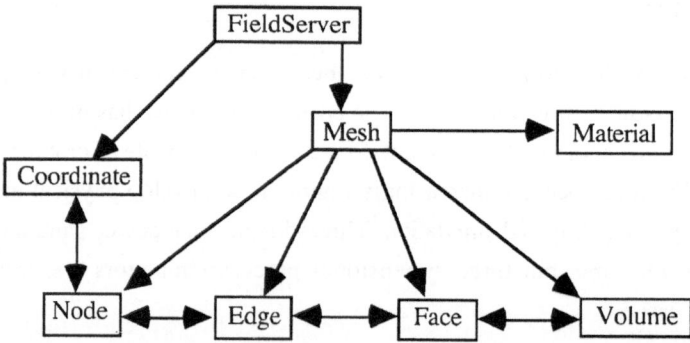

Figure 1 - Diagram showing the relationship of data objects in the grid. The element forms the base class of all the pieces of a mesh.

Each Mesh object is made up of a pointer to a material, a list of nodes, edges, and faces, and volumes. Each mesh all contains a list of boundary elements, which provides for internal holes. One of these boundary elements is the outer boundary and the remainder represent internal holes. Each mesh corresponds to a single material, and multiple meshes may exist with the same material. Each Node object contains a pointer to the single coordinate that represents the physical location of the node. The node also contains a list of all the incident edges. These edges are ordered, so that the near and far end can be determined uniquely. An edge in the mesh is made up of two nodes, and a list of all incident faces. A face contains pointers to three or more edges, each of which is ordered in a counterclockwise fashion. Each face contains pointers to all incident volumes. Volumes are made up of at least four faces.

All mesh objects contain pointers to the objects of one greater and one lesser dimension. The pointers to the objects of one lower dimension define the object, and the pointers to objects of one greater dimension are used to maintain traversal routines. For example, an edge contains pointers to two nodes, and to a list of faces incident on the edge. In a one-dimensional simulation, an edge has an empty set of faces incident. There is no direct list of nodes contained in a face, although the information is available through the ordered edges.

All the mesh objects (Node, Edge, Face, and Volume) are derived from the same base class, the Element. This simplifies development of a multi-dimensional process simulator, since any algorithm written to use elements will work in any dimension. In development of assembly and moving boundary sections of the code, elemental methods were used as much as possible. Using this technique, many of the algorithms used for three dimensions can be tested in two dimensions first. This simplifies the development of three-dimensional algorithms, since two-dimensional debugging is simpler.

The element derivation scheme allows simplified assembly as well. The assembly functions for each different partial differential equation take an element as an input, and produce as output the dense element Jacobian terms. Assembly is handled in two different ways depending on whether finite volumes or finite elements are being used to discretize the equations.

For finite elements, the assembly method is called for each element in the mesh. The assembly methods are used using dense matrix multiplies. The physics can be implemented in the same way, and the only difference is in the shape functions. The shape functions are coded for each different element type and discretization level. The two-dimensional and three-dimensional codes share the same physics, so debugging is simplified.

For finite volumes, assembly is always performed on nodes and edges. Flux terms are assembled on the edges, and recombination and time dependence are assembled on nodes.

The coupling coefficients and volume weightings are precomputed before assembly. Since the assembly is always performed on the same element types, the only changes required for three-dimensions is in the precomputation of the couplings and area weightings.

Grid Subtraction

During oxidation, mesh nodes need to be removed in front of the oxidizing interface. As the interface advances, tests are applied to all edges, faces, and volumes. For each of these element types, an effective shortest length is computed. For an edge, it is just the length of the edge. For a face, the effective shortest edge length is computed. This length is twice the triangle area divided by the length of the longest edge. This edge length is the minimum spanning distance across the triangle. Volumes are checked in a similar way. The minimum span is computed by using the maximum face area and the volume of the element. The last two lengths handle cases when the element is collapsing on itself and approaching zero area or volume.

Each of these lengths is checked using the same criteria. First, all lengths are checked to see if they are shrinking. If it is shrinking, the edge is removed if it has lost more than user-specified tolerance fraction of its length. Any edge smaller than a minimum threshold, usually 20Å, is always removed. For an element that spans a region, it is removed only if it is smaller than the minimum threshold. This insures that regions are not unduly distorted by the removal process.

Any edge that meets the above criteria is removed. If the edge is an effective edge, the edge is created first by splitting the face or the volume. By performing this operation, all removal algorithms work on edges only. Each node at the end of the edge is checked to see if it is internal to a mesh. The simplest internal nodes are those are not on any material border, but they can also be defined for nodes that are on contiguous material boundaries. If either end is internal, than that node can be removed with an internal reconnection step. If both ends are non-internal, *i.e.*, an edge that spans a material, a more complicated transfer algorithm is used.

Internal-like nodes are easily removed. An element is built which contains all the elements incident on the node. Because the element is the base class for all dimensions, the same code is used in two and three dimensions. The original incident elements are deleted, which leaves only the containing element in the area originally made up of all incident elements. This remaining element is divided into simplex elements. In two dimensions, this can be done without adding any interior nodes using a Delauney triangulator. In three dimensions, this is more difficult, as in general, nodes must be added to retetrahedralize the surrounding volume. We attempt to control this by adding points on the boundary rather than in interior, and only taking this stop when required. Figure 2 shows a sample case before and after removal in two dimensions. In short, this algorithm deletes the node and then retriangulates the area surrounding it.

Before:

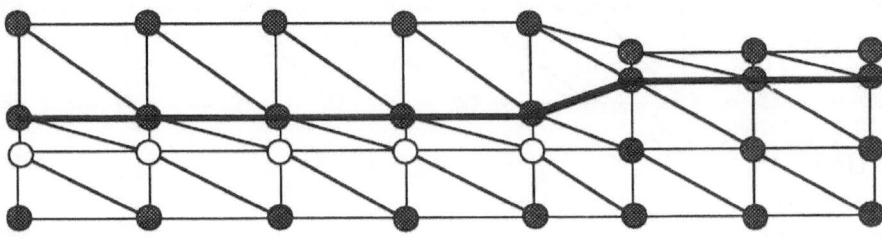

After:

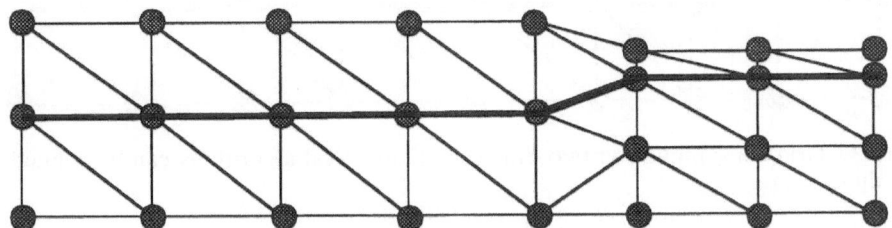

Figure 2 - Grid unrefinement near a moving boundary in two-dimensions. The white nodes in the top figure are removed by retriangulation of the surrounding area. A similar algorithm is applied in three-dimensions..

Edges that have non-internal nodes at both ends are more difficult to remove. In this case the surrounding elements are again computed, and they are transferred to the growing material. After element transfer, the existing region is checked to see if it has been split into two new regions. A single element object is marked, and all those objects that border it are also marked. This is implemented recursively so all objects that are attached are marked. Any unmarked objects left in the mesh are not contiguous with the remainder and must be put into a second mesh region. Some lower dimensional mesh pieces may be shared and would have to be duplicated, but this is easy to check and repair. At the end of this procedure, the original region will be split into two new regions. This insures that each mesh is contiguous.

Before:

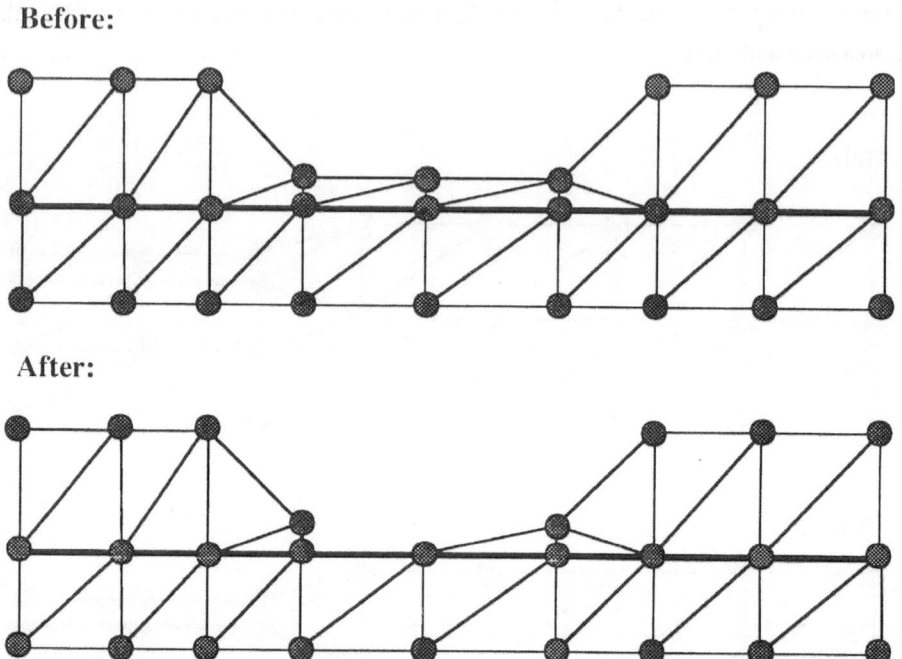

After:

Figure 3 - Grid consumption in two-dimensions. Identical algorithms can be applied in three-dimensions.

For example, Figure 3 shows a case where etching has almost completely removed a material layer. The node indicated has all of its incident elements transferred to the growing material (gas). The original elements are deleted from the mesh. In the case of oxidation, *e.g.*, through poly, the oxide layer will have grown incrementally. Because the removal criterion for this type of edge is typically very small, significant distortion of the materials is avoided. There can be some asymmetry in the removal depending on how the triangles are connected.

There are no special checks used to see if the surface is colliding[7]. This is because the gas is gridded in the same way other materials are. The gas grid can then be checked in the same way as all other meshes, and this handles surface collision problems.

Grid Addition

Grid addition is needed to resolve both the surface curvature and to refine the area behind the growing boundary. Ideally, refinement in all materials would be determined by error estimators of the partial differential equations. This approach has been used successfully in two-dimensions[8] for diffusion simulation, but it still relies on primitives to update the grid. Similar refinement primitives are needed for use with surface refinement for topography simulation[9, 10].

In FLOOPS, grid is generally stationary with respect to the material. In the case of oxide, for example, new material is formed at the boundary, and the old oxide is lifted. Old oxide nodes are lifted and the new volume created is associated with the silicon / oxide boundary nodes. Therefore, only the layer of grid adjacent to the boundary needs to be checked. An additional bonus is that dopants in the oxide move with the material; up and away from the interface, just as in a real growth process. A two part procedure is used in this algorithm. First, a list of candidate edges to be checked is produced. Second, these edges are checked to see if their current length is greater than the desired grid spacing.

Generating the candidate edges is straightforward. First, all edges that form the boundary of the mesh region are included. Then each surface point is checked to find the edge most perpendicular to the boundary at that point. This makes sure that new nodes are created perpendicular to the boundary, and that each interface node can have at most one new node added. This step makes sure that when grid is added to the interior, only a single point is added for each boundary point. This prevents over-refinement parallel to the boundary.

This sequence of edges is then checked for unacceptably long length. An edge is refined when its length exceeds the desired grid spacing. When an edge fails this test, it is refined. Refinement occurs at the midpoint, which we have found produces better shaped elements and improves convergence of the nonlinear flow solutions used in oxidation simulation[11]. The splitting procedure inserts an additional node in each of the elements that border the edge. These elements are then divided into simplexes, using one of a number of available algorithms[12, 13].

Grid Smoothing

To reduce the number of removals and grid related limits on the time step, a velocity relaxation is applied to the materials[14, 15]. This technique solves a Laplacian in the silicon and oxide to compute new nodal velocities. In the silicon, nodes are removed away from the interface. The nodes near the interface move faster than those farther away. Near surface edges don't shrink as fast, and therefore removals can be postponed. In the oxide, the silicon / oxide interface is moving down, and the oxide / gas interface is moving up. The nodes tend to move towards positions that distribute them equally through the material. Addition can now be performed less frequently, and good quality elements can be maintained. Figure 4 shows a two-dimensional result using smoothing away from the interface. The grid lines bend along the oxide interface.

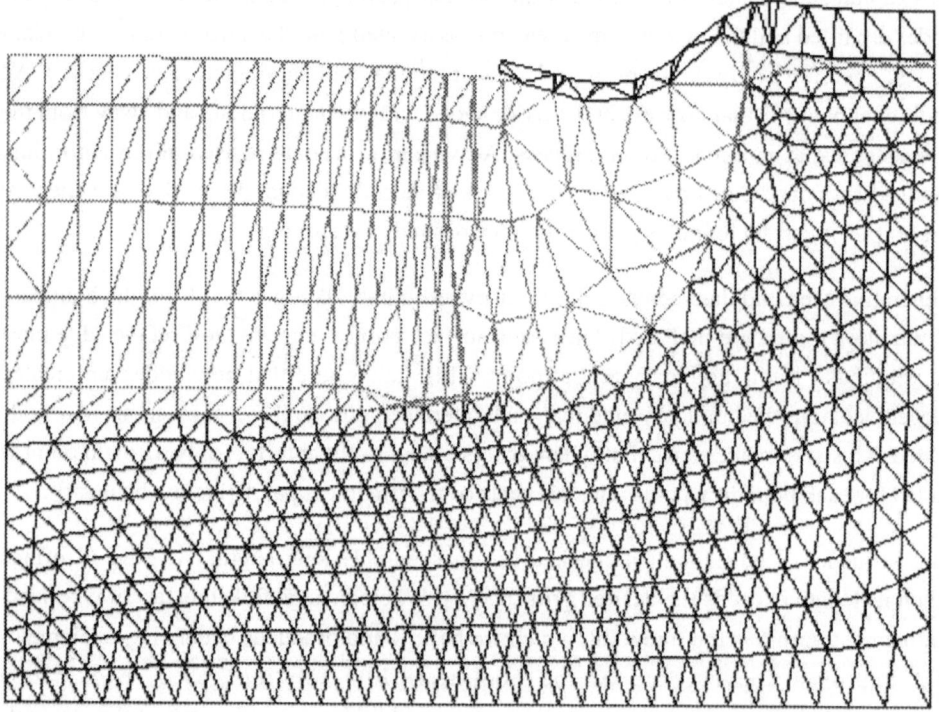

Figure 4 - Smoothing algorithm to move edges away from a moving boundary. The same algorithm is used in both two- and three-dimensions.

The nodal motion needs to be accounted for in the solution of the equations. For the diffusion equation solutions, the nodal velocity can be accounted fro by using an upwinding approach that subtracts the apparent diffusion from the point motion. This approach tends to introduce only small errors in the final simulations. The oxidation flow equations are easy when using a linear or nonlinear viscous model. In this case, the material flow has no dependence on the previous state and the grid is free to move. In the case of a viscoelastic model, the elasticity depends on the previous stress levels. In this case, moving the nodes can introduce error into the simulation. We do not know how to address this at present except with a simple smoothing algorithm.

Computation Issues

Figure 5 shows the CPU time required to compute an implant in two and three dimensions. The sample implant was 40KeV Arsenic at a dose of 10^{14} / cm^{-2}. The simulation sample was a rectangular mask opening in three dimensions, and a slice through this sample for two dimensions. The implant was computed using a single Pearson distribution with a constant lateral standard deviation. Dual Pearsons and depth dependent standard deviation models require more time to integrate across the surface.

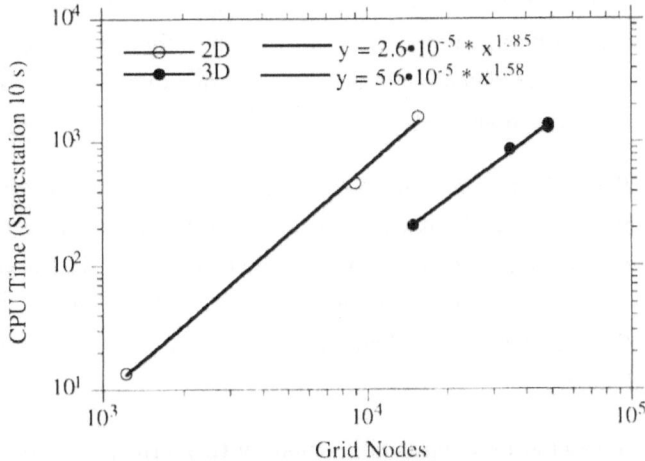

Figure 5 - Performance of implantation code using numerical integration across surface patches. The code has a strong non-linear dependence on dimension.

The implantation time is higher nonlinear in the nodes. Each surface patch must be implanted into and the contribution of each surface patch must be added together at each substrate node. The simplest implementation of the algorithm would require an iteration over all substrate nodes for each surface patch. This would produce an algorithm of complexity approximately $n \cdot n^x$, where n is the number of bulk nodes and n^x is a rough estimate of the surface nodes. In three dimensions, x would be roughly 2/3 and in two dimensions in would be roughly 1/2. Figure 4 indicates that this simple analysis is approximately correct.

In our case, we have obtained performance improvements through preprocessing of surface patches to find planar sections. Planar sections are unioned and treated as a single larger surface patch. This algorithm is roughly square in the number of surface nodes, but reduces the implant itself to something more linear in the number of bulk nodes. Figure 5 demonstrates this effect, as the savings is relatively greater in three dimensions than two. For the two-dimensional examples in Figure 5, the preprocessing of surface patches was the dominant time.

Further improvements will be required in the performance of the implantation. We currently use an adaptive Simpson's rule for surface integration of the point response, when the analytic integration of the doping is no longer appropriate. We might need to make larger use of the analytic formula and explore the accuracy gain / loss for the numeric integration. Additional performance could be obtained by preprocessing bulk nodes to determine if any reasonable dose contribution can be expected from a given surface patch. Similarly, nodes that are under a large surface patch could be given the one-dimensional value without integrating in the surface directions.

Figure 6 indicates the CPU time required for a single Newton iteration for an 1100°C anneal of the implanted structure discussed previously. The simulation was performed with the simplest diffusion model. No point defects were included in the simulation. The only solution variables were the arsenic concentration and the potential. Figure 6 indicates that the solution time is linear in the number of nodes, and that there is little difference between two dimensions and three.

The CPU time for a Newton iteration is dominated by two terms, assembly and matrix inversion. The assembly time is nearly linear in the number of nodes. Since finite volume techniques are used to discretize the partial differential equations, a single pass is made over edges and another on nodes. The number of edges is roughly a constant times the number of nodes for each dimension. A dimensional dependence is present in this term, since the edge count is greater in three dimensions than two. Figure 6 indicates that this is quite weak.

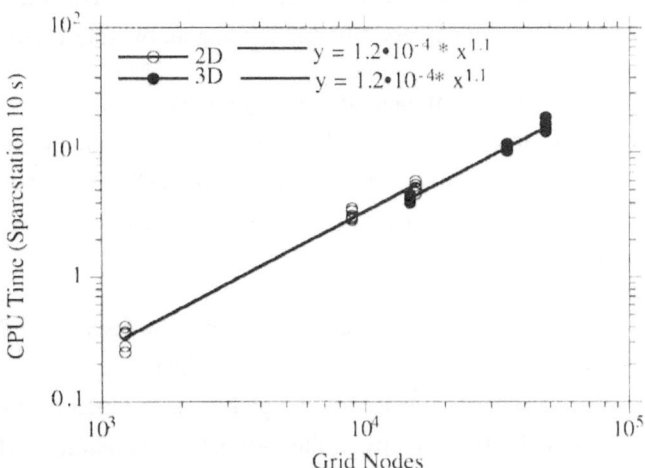

Figure 6 - Near linear dependence of diffusion CPU time on grid and dimension.

Matrix inversion is usually nonlinear in the number of nodes. In the case of diffusion, however, effective precondition iterative methods are available. FLOOPS uses an incomplete factorization and biconjugate gradient squared technique. This works extremely well, and has not failed to converge in any examples we have tried. The technique is so effective that assembly of the equations usually is the dominant CPU contribution. Further improvements in CPU time will probably come from more efficient assembly of the diffusion equations.

The total time for anneals depends on the number of time steps and the number of Newton iterations per time step. Our experience indicates that between five and fourteen Newton iterations as required per time step, depending on the nonlinearity of the problem. Heavier doping is more nonlinear, as are more complicated defect - dopant models. The number of Newton iterations does not depend on the number of nodes. The time step is TRBDF[16-18], and performs very well for most problems. The number of time steps depends on the anneal and models chosen. We have also noticed that the time step depends weakly on the density of grid points, and this could create a slightly larger dependence on node count for the total anneal CPU time.

Our initial experience leads us to believe that the simpler diffusion models will take less time for the anneal step then for the implantation step. In several of these examples, the implant step took more total CPU time than the anneal step. This will come as a shock to most users accustomed to something very different for smaller two-dimensional problems.

For oxidation, the CPU time is dominated by the time it takes to assemble and solve the flow equations. As we've already discussed, the solution of the oxidant diffusion equation is linear in nodes and contributes only a small fraction of the CPU time. The flow equation solution must be computed at least once per time step for linear models, and must be repeated for nonlinear models[11, 19].

Figure 7 shows the CPU time required to assemble and solve the discretized flow equations once for two- and three-dimensional simulation. The CPU time has a strong dependence on dimension. The number of equations is used as the plot variable rather than the number of nodes since an extra variable (the velocity in the z-direction) is required when changing dimensions.

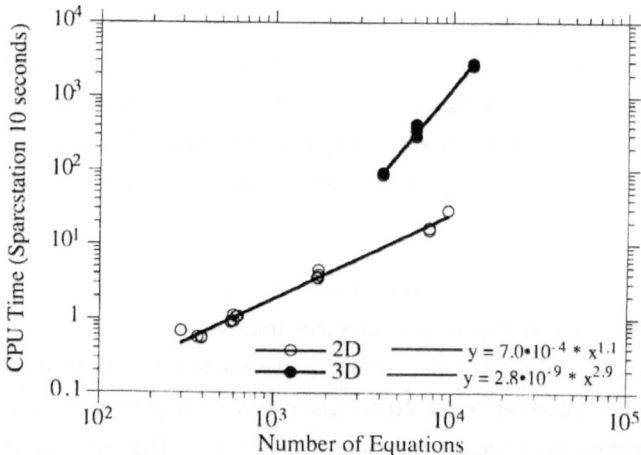

Figure 7 - CPU performance of a linear oxidation code in multiple dimensions.

The performance in two-dimensions is quite good. It approaches quite closely the performance of the diffusion algorithm, even though we have had little success with preconditioned matrix techniques. A full LU decomposition is used[20, 21] to solve the equations, but the dependence in this regime is approximately linear. The finite element discretization uses small dense matrix operations to perform assembly, and this can be the dominant time for smaller problems. The dense multiplies have a square dependence on the number of equations per element.

In three-dimensions, the performance deteriorates rapidly. The element matrices are much larger (since there are both more nodes and more equations per node), and the overall matrix is much more dense. The assembly and factorization times both become much greater. The nonlinear dependence is particularly disturbing, and will require significant work. Further work will be required on preconditioned iterative techniques for three-dimensional flow matrices. We also need to closely examine our assembly to attempt to improve performance of the dense matrix kernels used.

Computational issues in going from two to three dimensions appear mixed. Implantation starts to become a significant concern, when in two-dimensions it had negligible impact on the overall simulation time. Diffusion simulation appears to be well-behaved, and tends to remain linear in the number of nodes. Oxidation simulation appears to be quite CPU intensive compared to diffusion and implantation. Fortunately, the oxidation solutions need to be performed only on nodes above the substrate and not in the entire structure. This can create a substantial savings since the densest grid is required for resoling the dopant profiles. If stress is being computed in the silicon, it appears to be advantageous to decouple the silicon stress computation from the oxide flow solution. Oxidation simulation could easily be the largest contribution to the overall CPU time.

Examples

Figure 8 shows a perspective plot of a three-dimensional implant / diffusion simulation. A 0.5μm wide L-shaped mask was used for the implant, but stripped off so the details of the simulation could be seen more clearly. The implant was a 50KeV arsenic implant at a dose of $2 \cdot 10^{14}$ cm^{-2}. The anneal was 30 minutes at 950°C, with no point defects included in the simulation. The only diffusion variables are the arsenic concentration and the potential.

The structure has 30,000 nodes. The implant took approximately 1000s of CPU time, and the anneal required 1200s of CPU time. The simulation was performed using bricks as the discretization elements. The simulation shows the expected result of a reduced penetration of the dopant in the corner of the mask due to the limited dopant source available. The most interesting component of the simulation is that the implant takes nearly as much time as the diffusion simulation. Point defect based simulations would take more time, since the structure would need to be larger to include the defect migration appropriately, and since it would increase the number of variables in the matrix.

Figure 8 - Three-dimensional implantation and anneal into the corner of mask. The doping contours are in gray-scale. The mask opening was 0.5μm on a side.

Figure 9 shows a three-dimensional simulation of a standard LOCOS structure using nonlinear viscoelastic models for the growth[11]. The nitride mask has been stripped off so the stress contours can be seen in the oxide. The growth temperature was 1000° C LOCOS and the nitride mask was 0.6um by 2um and 1000Å thick. The final field oxide thickness is ~3500Å. The simulation was performed with bricks and used velocity relaxation as the primary technique for grid adaption. The oxide simulation indicates that this structure is not two-dimensional at any place. The bird's beak length parallel to the narrow edge of the nitride is ~0.35um and along the longer edge is ~0.2um. It appears that the three-dimensional stress profile is influencing the growth of the oxide and the formation of the bird's beak.

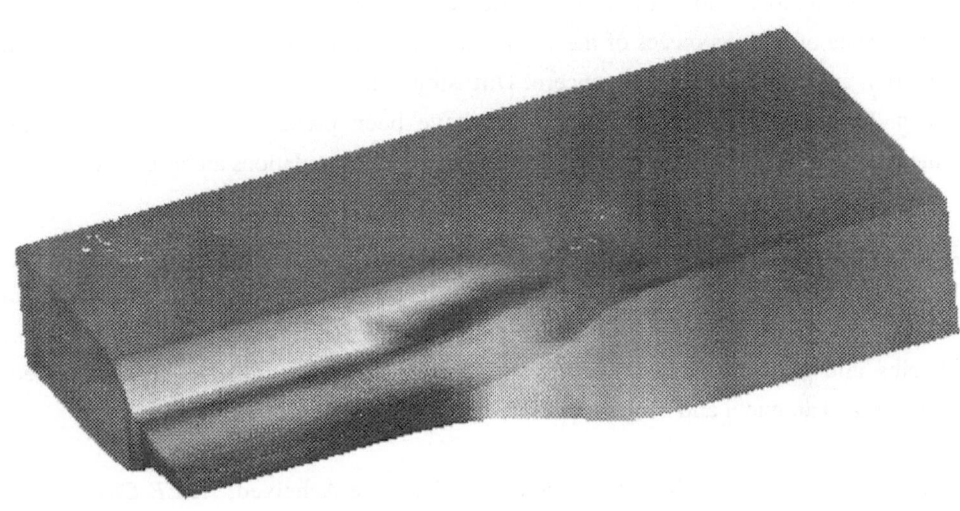

Figure 9 - Three-dimensional oxidation of a standard LOCOS shape. White regions are in tension of $1.5 \cdot 10^{10}$ dynes/cm^2 and the dark regions are in compression of $4 \cdot 10^{10}$ dynes/cm^2.

As the nitride is bent along the longer dimension, it creates a region of compression in the nitride toward the middle of the stripe. Figure 9 shows the compressive region in the edge of the pad oxide near the middle of the oxide stripe. This region of compression makes the nitride seem stiffer in the narrow direction, and creates less encroachment in that direction. The higher tensile regions (white) are further under the tip of the stripe indicating the increased growth and lifting of the nitride at the long end.

Conclusions

We have developed a three-dimensional capability in the process simulator FLOOPS for oxidation, implantation, and diffusion. The key challenge of grid motion has been met by exploiting the object-oriented approach used in FLOOPS. We've been able to extend the two-dimensional methods for grid to three-dimensions in a simple way. The CPU performance of various pieces of the simulator has been compared, and implantation and oxidation appear to be a major concern. Diffusion simulation is linear in the number of nodes. Examples of the new capabilities have been presented, including a three-dimensional oxidation that indicates that two-dimensional simulations are insufficient.

References

1.   C. Hegarty, T. Feudel, N. Hitschfeld, R. Ryter, N. Strecker, M. Westermann and W. Fichtner, "An Approach to Three Dimensional VLSI Process Simulation," Process Physics and Modeling in Semiconductor Technology Symposium, Honolulu, ed. Srinavasan, Taniguchi and Murthy, p. 565, 1993.

2.   M.E. Law, "The Virtual IC Factory ... Can it be Acheived," *IEEE Circuits and Devices*, **11**(2), p. 25, 1995.

3.   M.E. Law, "Challenges for Achieving Accurate Three-Dimensional Process Simulation," Simulation Semiconductor Devices and Processes, Vienna, ed. Selberherr, Stippel and Strasser, p. 1, 1993.

4.   M.D. Giles, D.S. Boning, G.R. Chin, W.C. Dietrich, M.S. Karasick, M.E. Law, P.K. Mozumder, L.R. Nackman, V.T. Rajan, D.M.H. Walker, R.H. Wang and A.S. Wong, "Semiconductor Wafer Representation for TCAD," *IEEE Transactions on CAD*, **13**(1), p. 82, 1994.

5.   M. Liang and M.E. Law, "An Object-Oriented Approach to Device Simulation - FLOODS," *IEEE Trans. on CAD*, **13**(10), p. 1235, 1994.

6.   M.E. Law, C.S. Rafferty and R.W. Dutton, SUPREM-IV User's Manual, 1988.

7.   J.J. Helmsen, E.W. Sheckler, A.R. Neureuther and C.H. Sequin, "An Efficient Loop Detection and Removal Algorithm for 3D Surface-Based Lithography Simulation," NUPAD, Seattle, p. 3, 1992.

8.   C.C. Lin and M.E. Law, "Mesh Adaption and Flux Discretizations for Dopant Diffusion Modeling," NUPAD-V, Honolulu, p. 151, 1994.

9.   E.W. Scheckler and A.R. Neureuther, "Models and Algorithms for Three-Dimensional Topography Simulation with SAMPLE-3D," *IEEE Transactions on Computer-Aided Design*, **13**, 1994.

10.   H. Rueda and M.E. Law, "Algorithms for Reduction of Surface Evolution Discretization Error," SISDEP, Erlanged, 1995.

11.   S.M. Cea and M.E. Law, "Multi-Dimensional Nonlinear Viscoelastic Oxidation Modeling," SISDEP, Erlanged, 1995.

12.   C.L. Lawson, "Software for C^1 Surface Interpolation," Mathematical Software III, Rice ed., Academic Press, New York, 161, 1977.

13.   B. Joe and R.B. Simpson, "Triangular Meshes for Regions of Complicated Shape," *Intern. J. Num. Meth. Eng.*, **23**, p. 751, 1986.

14.   K. Smith and R.E. Bank, "Optimal Moving Meshes for Process and Device Simulation," NUPAD, Seattle, p. 187, 1992.

15.   M.R. Pinto, D.M. Boulin, C.S. Rafferty, R.K. Smith, W.M. Coughran, I.C. Kizilyalli and M.J. Thoma, "Three-Dimensional Characterization of Bipolar Transistors in a Submicron BiCMOS Technology Using Integrated Process and Device Simulation," International Electron Devices Meeting, San Francisco, p. 923, 1992.

16.   R.E. Bank, W.M.C. Jr., W. Fichtner, E.H. Grosse, D.J. Rose and R.K. Smith, "Transient Simulation of Silicon Devices and Circuits," *IEEE Trans. on Electron Devices*, **ED-32**, p. 1992, 1985.

17.   H.R. Yeager and R.W. Dutton, "An Approach to Solving Multi-Particle Diffusion Exhibiting Nonlinear Stiff Coupling," *IEEE Trans. on Elec. Dev.*, **32**(10), p. 1964, 1985.

18.   M.E. Law and R.W. Dutton, "Verification of Analytic Point Defect Models using SUPREM-IV," *IEEE Trans. on CAD*, **7**(2), p. 181, 1988.

19.   D. Collard, B. Baccus and B. Hamonic, "A Robust Numerical Procedure for Stress Dependent 2D-Oxidation Similation," NUPAD-IV, Seattle, 1992.

20.   T.A. Davis, Performance of an Unsymmetric-Pattern Multifrontal Method for Sparse LU Factorization, 1992.

21.   T.A. Davis, Users' Guide for the Unsymmetric-Pattern MultiFrontal Package (UMFPACK), 1993.

3D Process Simulation at IEMN/ISEN

B. Baccus, S. Bozek, V. Senez and Z.Z. Wang

IEMN Département ISEN,
41, Boulevard Vauban, 59046 Lille Cédex, FRANCE

Abstract

This paper addresses the current fields of interest at IEMN/ISEN concerning 3D process simulation. The emphasis is on the diffusion and oxidation steps and the associated issue of 3D mesh generation. For each step, the achievements are presented with special attention to the numerical aspects. In particular, the principles underlying local remeshing are discussed.

1. Introduction

It is now well recognized by the industrial users that 3D process simulation can be helpful for the development and analysis of advanced silicon technologies. The first application consists certainly in the generation of 3D doping profiles and geometries even in a 'simplified' manner, in order to be used as input for 3D device simulation. This can be achieved either by a multidimensional approach [1, 2] or by actual 3D process simulation with basic models. On the other hand, there are also intrinsic 3D effects which require a level of modelling similar to the ones used in 1D and 2D simulators [3]. While the implementation of the physical models is relatively straightforward, with the limitation of their use by the natural increase in CPU time with the number of unknowns, basic numerical problems appear when introducing the third dimension, whatever the level of modelling. This is especially clear for the oxidation step, for at least two reasons. First, the solution of a viscous or visco-elastic mechanical problem requires the inversion of a matrix generally ill-conditioned, which explains the large calculation times or even convergence problems in 2D. This can be only worst in 3D, due to the increase in the number of nodes and connectivity. Second, the moving boundary problem associated with the oxidation kinetics asks for the structures and mesh updates at each time step. Schemes have been proposed in 2D with various success, while very few has been reported for 3D. This leads naturally to the issue of mesh generation and adaptation in its wider acceptation, which undoubtedly is one of the main problem for 3D simulation [4].

This discussion brings us quite far from the user's final requirements, but it is fair to acknowledge that, although the TCAD community can take great benefit from the experience gained in 1D and 2D programs, basic numerical problems are still pending which will ask for large efforts in the coming years. Within the frame of the european PROMPT project, we are mainly interested in the 3D simulation of the diffusion and oxidation steps, and to some extent to the mesh generation and adaptation since it is difficult to separate this latter problem from the concerned processing steps. In section 2, the meshing strategy is presented, while section 3 and 4 describe the diffusion and oxidation modules, respectively.

2. Meshing strategy

Following our previous works on 2D simulation [5, 6], we have again adopted the finite element method (FEM) for the solution of the diffusion and oxidation equations. We shall then consider mesh generation from two points of view: a) the initial mesh and b) the update of the mesh during the oxidation step.

2.1. Initial mesh

The aim of the initial mesh is obviously to respect a defined geometry with a minimum number of nodes, taking eventually into account some doping distributions. For this purpose we have make use of the mesh generator Ω from ETH Zürich [7]. Although it was initially designed for device simulation, the present application to process simulation is straightforward. On the other hand, it generates meshes containing different types of elements: bricks, prisms, pyramids and tetrahedra. The FEM allows an easy implementation of the corresponding shape functions and our first version of the diffusion module uses all types of elements, as will be shown in the diffusion examples. However, from a general perspective, we have found desirable to use only tetrahedra, as will be explained in the next subsection. We then use the latest version of Ω, which generates a tetrahedral mesh from the mixed-elements grid and subsequently constructs a Delaunay mesh by geometrical transformations. On the other hand, the underlying concepts of Ω prevent its use for local remeshing during the updates after each oxidation step. Since it is not reasonable to rely on an approach which would consider a complete remeshing, an other strategy has been adopted and is presented below.

2.2. Requirements for local remeshing

There are many requirements for local remeshing, especially when considering its application to the moving boundary problem.

First, the geometries are generally extremely complex (e.g. double-poly bipolar devices, EEPROM cell), which ultimately limit the possible meshing techniques. More precisely, the structures are most of the time non-convex, eventually multiply connected, include very thin layers, and of course contain local refinements. Let's now imagine a complete simulation structure as the union of non-overlapping domains determined by the envelopes consisting of triangular faces. Then, the mesh built within each domain should respect these specified boundaries (envelopes), or in other words the mesh is constrained.

2.3. Delaunay mesh generation

Among the several techniques proposed for 3D mesh generation, the so-called Delaunay (or Voronoï) methods seem very promising [8]. Very successful results have been obtained by different groups in 2D process simulation codes, and this method is theoretically extendable to 3D. As it makes use of simplices in most cases, we shall first discuss their advantages and drawbacks for application to the present problem. Then the method will be briefly presented with some applications.

2.3.1. tetrahedral meshes

From a general point of view, it is well-known that it is *in principle* easier to fit any geometry by the use of the simplices (triangles in 2D and tetrahedra in 3D). In practice, this statement relies on the availability of a powerful mesh generator, which is not necessary the case, especially in 3D. Thus, one should better consider this as a wish ! From a more pragmatic point of view, it is certainly easier to manipulate tetrahedra during mesh (and thus element) deformation as might occur in oxidation. In the same manner, this argument holds for *local* mesh updates, where it might be difficult to consider bricks in very complicated geometries already containing other elements. Finally, the simplicity of the shape functions not only makes easier the writing of the assembly step, but also minimizes the associated calculation time.

Conversely, the use of tetrahedral meshes leads to a dramatic increase of the connectivity, i.e. the number of elements containing a given node. As a result, more non-zero terms appear in the final matrices and hence the convergence of the (iterative) linear solver is somewhat lowered. As a rule of thumb, a connectivity of 6 to 9 is observed in 2D, but is frequently on the order of 50 in 3D. The consequences for a usual diffusion problem are not very significant, while this asks for very efficient iterative solvers for the solution of the mechanical problem (oxidation). Obviously, the situation becomes intractable for direct solvers, due to the increase of the bandwidth and memory requirements. In addition, it is often stated that the mesh tends to be quite uniform and contains a very large number of nodes and elements. This is however not necessarily true and such an effect can be minimized, provided, again, that one can use a 'powerful' mesh generator.

2.3.2. the method

This is fundamentally a scalar process, since the points are successively inserted into an existing mesh, with the property that the obtained triangulation is of the Delaunay type. This means that the circumsphere associated with each tetrahedron does not include any other point. Basically, for each new point P to be inserted in the mesh, it must be determined whether P is located inside, outside or on the spheres determined by the existing tetrahedra. Then, the concerned tetrahedra are deleted, new elements being included. Although conceptually simple, the practical implementation is relatively difficult due to the rounding errors. Moreover, the CPU performance can vary very much depending on the specific implementation of the algorithms. A more detailed description can be found elsewhere [8, 9]. The respect of the specified boundaries is also an issue, although significant progress has been made in the past few years. In any case, this is of course related to the quality of envelope discretization, which fortunately can be to some extent controlled as a post-processing of the oxidation module delivering the new interfaces.

This method guarantees the best quality for a *given* set of points. In general, smoothing techniques are added to increase the quality of the elements. In relation with this latter problem, it should be noted that in contrary with common techniques for mesh refinement, the degradation of elements quality can be limited by the full use of this method. Indeed, once a mesh is available, it is generally refined by the introduction of mid-edge points. In 2D, most of the degradation occurs for the transition elements, while in 3D it is even not possible to subdivide a tetrahedron in similar elements. On the other hand, if the method of point insertion is fully applied as described above, including the deletion of the tetrahedra whose circumsphere includes that point, then the Delaunay criterion is automatically fulfilled and the quality can be controlled.

2.3.3. examples

The first example illustrates the application of the method to a non-convex structure (fig. 1). Although there is a description of the boundaries, the resulting mesh will be necessarily convex and will contain external elements (fig. 2). These are then removed to restore the desired mesh (fig. 3). It is possible to facilitate these steps by the introduction of a bounding box [8].

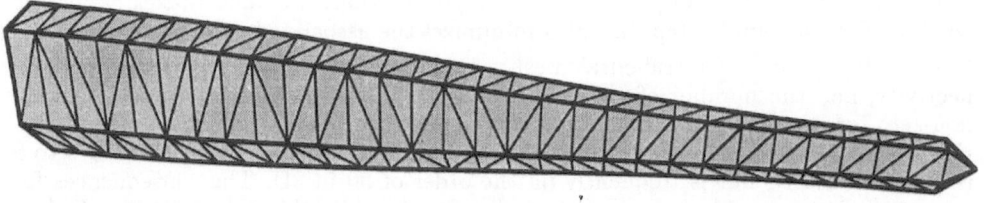

Figure 1: Initial definition of the structure by the surface

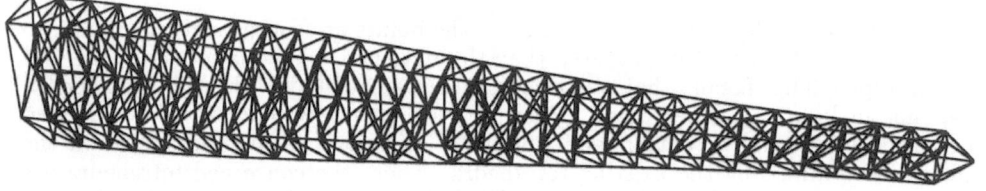

Figure 2: Delaunay mesh generation

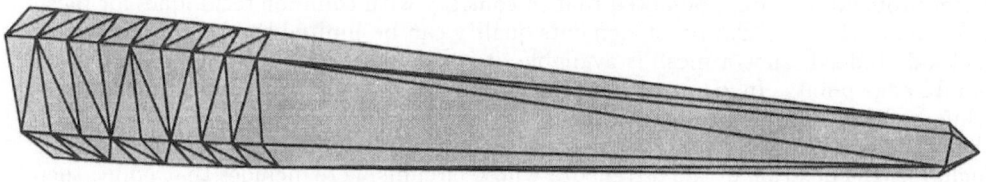

Figure 3: Final mesh after the removal of external elements

The second example shows a mesh refinement using the present method. A very simple criterion has been used, i.e. the ratio of concentrations between nodes of each element. The mesh is then regenerated as described in the previous subsection. The structure contains 33,793 tetrahedra and 5,192 nodes (fig. 4).

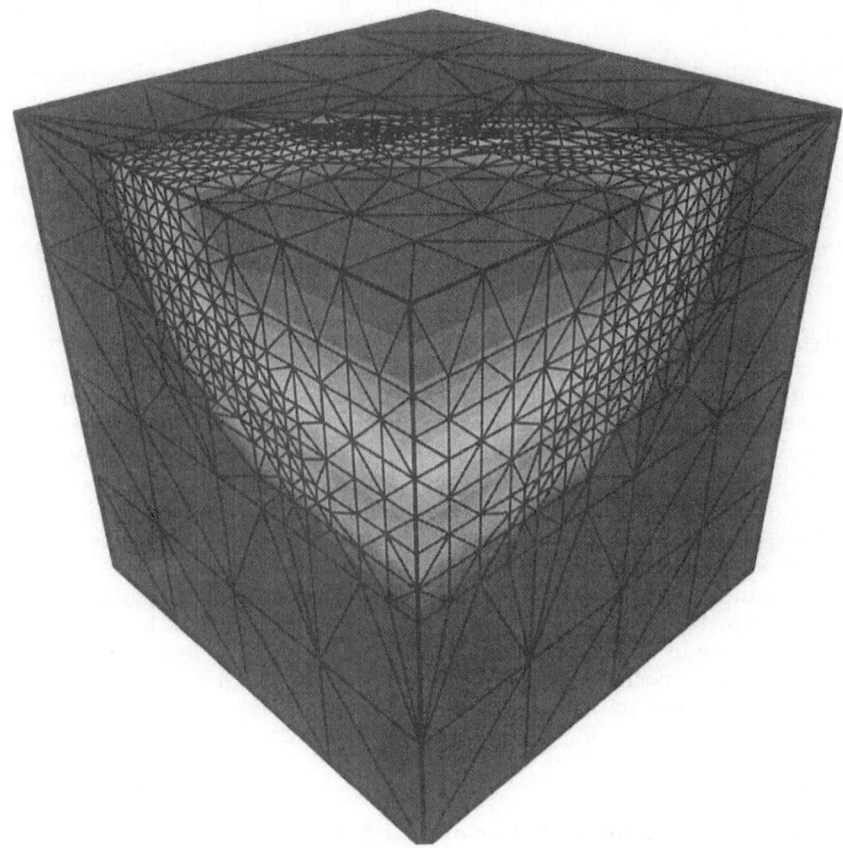

Figure 4: Example of mesh refinement

3. Simulation of the diffusion step

In this section, we present the present achievements in the simulation of diffusion steps (in inert ambient). The modelling level will be first described, followed by the numerical implementation by the FEM, examples on bipolar devices and then envisioned developments.

3.1. Modeling level

As a first stage, a conventional diffusion model has been chosen. The main motivations were the early deliverable of 3D profiles, as well as a minimal CPU time.

Obviously, it is well-known that point-defect based diffusion models are required for modern technologies and the future works will include such a formulation. In addition, it should be noted that it is still possible to include in an empirical manner the influence of point-defects through effective diffusivities, e.g in [10]. These are generally 1D formulations, and it is still an open question whether it is possible to introduce efficiently 2D or 3D effects. On the other hand, as stated in the introduction, the main problems to be addressed are primary of numerical type, even for simple physical formulations.

The model implemented is a conventional macroscopic model in which the point-defects are assumed to be at equilibrium [11]. This includes the influence of the different charge states of the dopant-defect pairs and the electric field term, allowing the study of coupled dopant diffusion. Additional features are a static clustering model for arsenic and boron in high-concentration and Fair-Tsai model for phosphorus diffusion [12]. In addition, the segregation between different materials is also considered, for example at the Si/SiO_2 and $Si/poly-Si$ interfaces. Classical values of the parameters have been used [13, 14].

3.2. Numerical implementation

This diffusion equation has been implemented by the use of the FEM. Linear shape functions are used, for all types of elements (bricks, prisms, pyramids and tetrahedra) in a first stage, while the following versions of this diffusion module will only make use of tetrahedra. The temporal discretization is performed by an incomplete implicit scheme with a constant time step. A mass-lumping technique is also used in order to enforce the stability of the linear system. It is solved by the conjugate gradient method with a basic diagonal preconditioning. While such a simple iterative solver is sufficient to obtain acceptable CPU times, further improvements could be obtained from more sophisticated algorithms. Calculations on test examples have shown an almost linear variation of CPU time with the number of nodes, at least up to 100,000 unknowns. Finally, a special scaling technique is employed for the simulation of segregation effects in order to obtain a symmetric matrix [5].

3.3. examples

The first example concerns the formation of a NPN bipolar transistor. The mesh is first generated by Ω and contains 75,345 elements and 77,057 nodes. The substrate is phosphorus-doped at a concentration of 10^{16} cm^{-3}. In order to define the base region, boron is predeposited through a base 'contact' which includes the intrinsic and extrinsic regions. Diffusion is performed at 900°C for 60 min, assuming a constant surface concentration of 5.10^{18} cm^{-3}. Finally, arsenic is predeposited through the emitter contact, with a constant surface concentration of 1.10^{20} cm^{-3}. This step is performed at 950°C for 30 min. During this latter stage, the three impurities are simultaneously diffusing and exhibit classical coupling effects. The fig. 5 shows the corresponding arsenic contours, demonstrating a very good stability. The boron contours can be seen in fig. 6 at the end of the simulation. Retarded diffusion in the emitter region is clearly observed, while the classical dip due to the electric field term is also present. Finally, the fig. 7 displays the net contours.

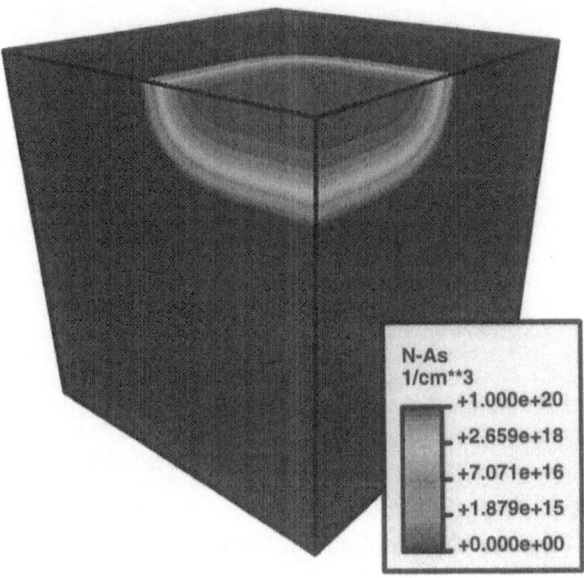

Figure 5: Arsenic contours in the bipolar transistor

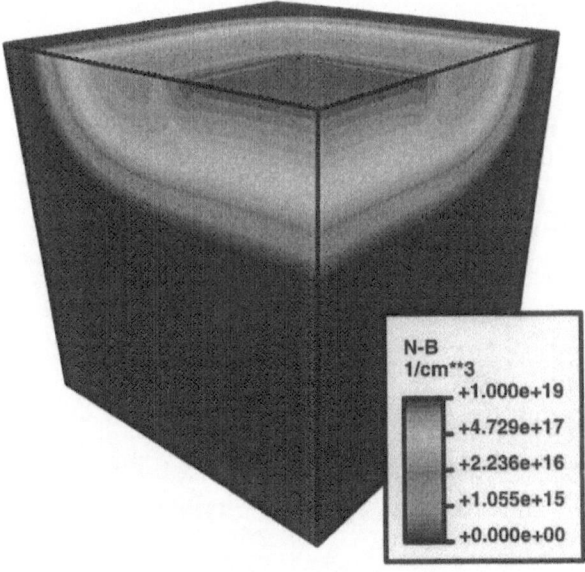

Figure 6: Boron contours in the bipolar transistor

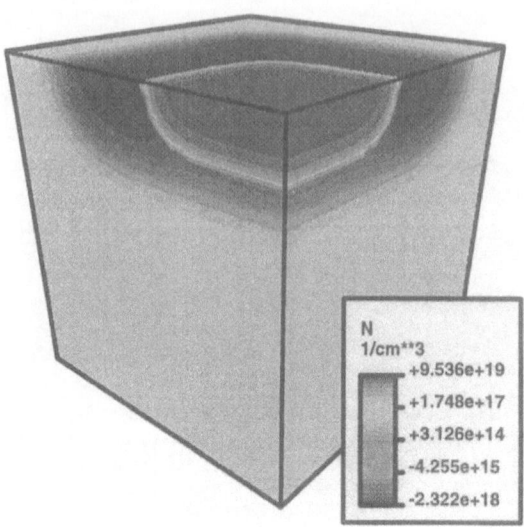

Figure 7: Net contours in the bipolar transistor

The second example illustrates the simulation of the segregation effect in a Si/SiO_2 structure. Again, a predeposition step is performed, in the present case with arsenic at a surface concentration of 10^{18} cm^{-3}, at 1000°C for 30 min. Due to the lower diffusion coefficient in the oxide, the oxide layers can act as masks, while the arsenic diffusion in silicon induces the segregation effect at the Si/SiO_2 interface (fig. 8). The structure includes 17,344 elements and 18,831 nodes.

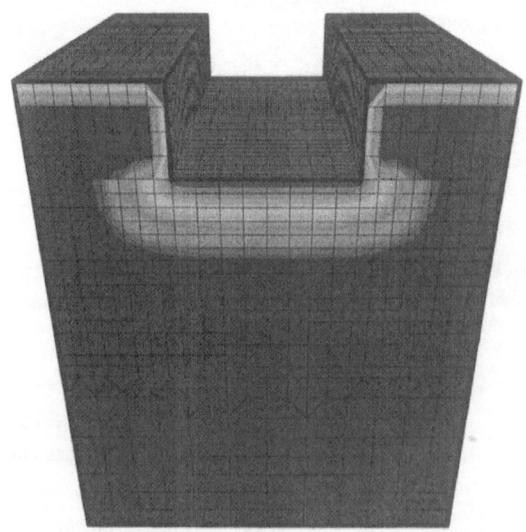

Figure 8: Arsenic diffusion and segregation in a Si/SiO_2 test structure

3.4. future works

The future works will obviously have the aim of improving the modelling level. At first, the introduction of a point-defect model is mandatory, it order to be able to treat actual 3D effects, like the effect of source/drain implants on the channel profile in the width direction, which can not be treated by 2D simulators. In the same spirit, while the present version makes use of an empirical approach for the diffusion in polysilicon through effective diffusivities, it is intended to implement a much more elaborate formulation, in which the dopant diffusion and grain growth are physically formulated [15]. Finally, it is useful to integrate advanced models only if one can be absolutely confident on the numerical accuracy. In 3D, this is a real issue, since it relies on the availability of efficient mesh adaptation algorithms. This comprises not only the remeshing procedure as described in the preceding section, but also the criteria to be used [16] and eventually higher order discretizations [17].

4. Simulation of the oxidation step

We shall use in this section the same presentation format than for the simulation of the diffusion step, but noting that the numerical problems are much more severe. In particular, some solutions will be proposed for the remeshing step after each oxidation step.

4.1. Modeling level and numerical implementation

Following our previous works on 2D oxidation [6], a visco-elastic formulation has been chosen which allows the realistic simulation of a large number of structures for very different processing conditions. Moreover, a better convergence has been observed, compared to a viscous approach. As a first stage, a linear model has been implemented, which does not include the influence of stress on the oxidation kinetics (diffusion coefficient, reaction rate, viscosity). Again, such an approach does not preclude the introduction of more sophisticated effects, but it enables a progressive treatment of the many numerical difficulties within a reasonable CPU time.

The first step requires the simulation of the diffusion of oxidants through the existing oxide, in the numerical form of the Laplace equation. In a second step, the mechanical deformation is calculated by a visco-elastic formulation : both silicon dioxide and nitride are considered as viscoelastic materials, with the Maxwellian formulation. On the other hand, the silicon and polysilicon layers are treated as rigid bodies. In the simulation, and due to the temporal discretization, the strain velocity is kept constant during each time step, which allows the use of an effective modulus of rigidity [18].

Although the formulation is presently in a linear form, special care must be taken for its numerical solution. The mechanical problem is solved by the FEM, using linear shape functions on tetrahedra. As described previously, the convergence of the linear system is degraded due to the large connectivity. We have found sufficient in the present case to use a standard conjugate gradient method, with however a large sensitivity of the convergence rate on the values of the parameters, e.g. viscosities. This is coherent with our 2D observations, where a direct solver has been used [6]. The solution of the linear system in the frame of the general visco-elastic modelling, including non-linear effects such as plasticity, is still an open question and will require significant efforts in the future.

4.2. examples

Three examples are reported below. In both cases, the silicon substrate is not displayed since the remeshing strategy is currently under development and is explained in the following subsection. The first example represents the oxidation of a LOCOS structure with a single nitride layer. The mesh is composed only of tetrahedra and is updated according to the displacements calculated at each time step. This is obviously a simplified approach, but it allows the first tests of the mechanical part of the problem. The structure at the end of the oxidation step at $1000°C$ for 90 min in wet ambient is displayed in fig. 9. The oxidant concentration in the oxide is also reported. The simulation includes 2,156 nodes and 9,306 elements. A cut within the structure is reported in fig. 10 and shows the distribution of oxidants in the corner direction.

Finally, the fig. 11 represents a similar structure, in which the nitride layer has been removed, displaying the oxidant concentrations at the SiO_2 surface. The FEM mesh as 42,648 nodes and 215,292 tetrahedra. In all these examples, one can see that the oxide slope is larger in the nitride corners, due to the lower concentrations of oxidants. Moreover, it is more difficulty to bend the nitride there. This 3D effect is even more pronounced when introducing non-linear stress effects [3].

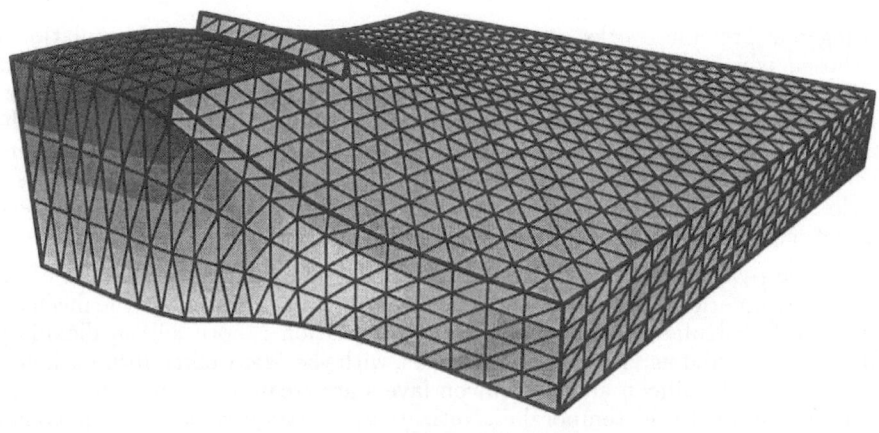

Figure 9: Oxidant concentration in a 3D LOCOS structure

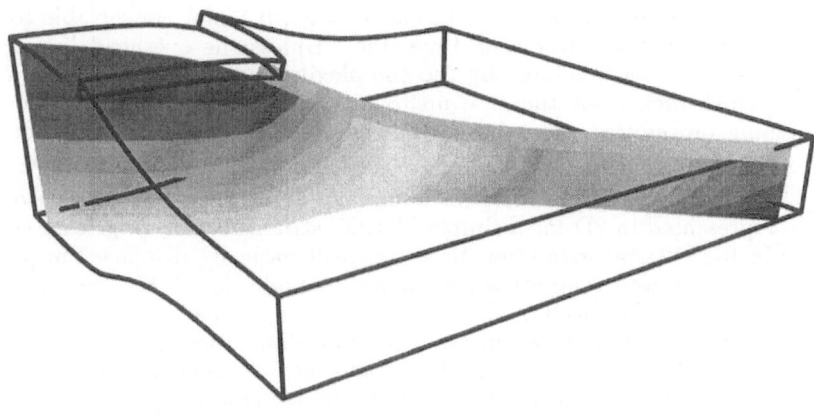

Figure 10: Cut along the corner of the preceding simulation result

Figure 11: Oxidant concentration in a 3D LOCOS structure. The nitride layer has been removed, but its limits corresponds to the line drawn on the LOCOS

4.3. regridding aspect

As discussed in the section on the meshing strategy, it is not conceivable to remesh completely the structure after each time step. Unless the overhead in calculation resources would be overwhelmed by the complexity and lack of reliability of other solutions (which seems not the case up to now), the present problem requires local remeshing procedures. This has to be compared with the 2D situation, where complete remeshing is still an efficient approach in terms of CPU time [6].

The fig. 12 summarizes an 'optimal' (or minimal!) scheme [9]. For the sake of clarity, it is here represented in 2D for a simple LOCOS structure. Starting from the initial structure in fig. 12a, an oxidation time step will basically displace the previously existing oxide layer and generate a new 'band' of oxide (fig. 12b). We shall assume that the mesh in the displaced oxide is still acceptable, or that its quality can be kept with the use of smoothing procedures. The main difficulties are anyhow to update the silicon mesh and connect the 'displaced' and 'new' Si/SiO_2 interfaces. The basic idea consists in performing a minimal remeshing procedure by considering the two following areas:

- the 'band' (shell in 3D) defined by the hull consisting of the two Si/SiO_2 interfaces ('displaced' and 'new'). This hull can be obtained quite easily, while the resulting geometry will be generally very complex in terms of aspect ratio and variation of the slopes.

- an other 'band' defined by the hull consisting of the 'new' Si/SiO_2 interface and a lower interface in the silicon bulk (not shown on the figure). This latter one can be determined by searching for the list of tetrahedra which intersect with the 'new' Si/SiO_2 interface and defining their opposite faces as part of the above-mentioned lower interface.

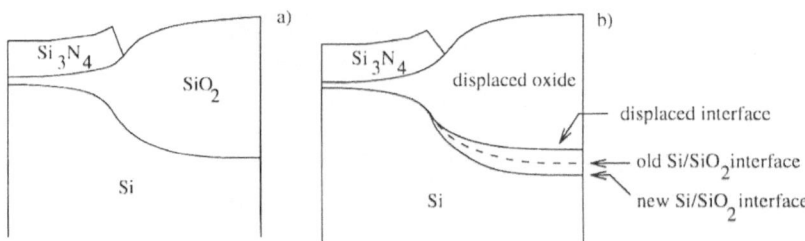

Figure 12: Remeshing strategy during oxidation, with a structure a) before, and b) after a time step

Such a scheme requires a powerful mesh generator, that basically respects the prescribed hulls. To our knowledge, the only technique which hopefully can solve this issue is the Delaunay method as described in the meshing strategy section. An example of the mesh generation in such a narrow band is reported in figs. 1-3, which supports the above analysis. This is of course an optimal scheme, and it is probable that after the sequence described here, smoothing techniques will be required in order to increase the element quality. It is also possible, once the different submeshes have been connected together, to introduce new points at that interfaces.

5. Conclusions

2D programs have really reach a stage of maturity after a significant period of time, and still, dramatic improvements have been observed very recently in the modeling aspect. Moreover, despite many efforts, mesh adaptation is a field of intensive research. This has to be compared with the 3D situation, where very few has been done in the past. In particular, the development of a *general* 3D grid generator is still a key and open problem. It is of interest to see that strategies defined for 3D device simulation can be of some help only to some extent, while original approaches must be investigated for the moving boundary aspect. A solution has been proposed here, based on a local remeshing technique and Delaunay criterion. This is obviously a long term approach and very basic problems remain to be solved. One should just remember that such approaches were contemplated a few years ago for 3D device simulation [19] but were discarded, due to the state of the art at that time. Unfortunately, such efforts are probably the price to pay in order to limit the calculation time and have enough confidence in the accuracy of simulations performed with up-to-date models.

6. Acknowledgments

This work is part of PROMPT (JESSI project BT8B) and was funded as ESPRIT project 8150. The authors would like to thanks ISE AG for the use of PICASSO and SIMBAD. Helpful discussions with P.L. George are also gratefully acknowledged.

References

[1] S. Ushio, K. Nishi, S. Kuroda, K. Kai and J. Ueda, "A Fast three-dimensional process simulator OPUS/3D with access to two-dimensional simulation results," *IEEE Trans. on Computer aided design*, vol. CAD-9, pp. 745-751, 1990.

[2] W. Fichtner et al, "Multidimensional TCAD: the PROMPT/DESSIS approach," *this workshop*.

[3] H. Umimoto and S. Odanaka, "Three-dimensional numerical simulation of local oxidation of silicon," *IEEE Trans. Electron Devices*, vol. ED-38, pp. 505-511, 1991.

[4] M.E. Law, "Challenges to Achieving Accurate Three-Dimensional Process Simulation," In Proc. *Simulation of Semiconductor Devices and Processes*, vol. 5, pp. 1-8, 1993.

[5] B. Baccus, D. Collard, E. Dubois and D. Morel, "IMPACT-4, a general two-dimensional multi-layer process simulator," In Proc. *Simulation of Semiconductor Devices and Processes*, Eds. G. Baccarani and M. Rudan, Vol. 3, pp. 255-266, 1988.

[6] V. Senez, D. Collard, P. Ferreira and B. Baccus, "Simulation of advanced field oxidation using calibrated viscoelastic stress analysis," In Proc. *IEDM'94 Conf.*, pp. 881-884, 1994.

[7] N. Hitschfeld, *Grid Generation for Three-Dimensional Non-Rectangular Semiconductor Devices* , Hartung-Gorre, PhD thesis, ETH Zürich. 1993

[8] P.L George, *Automatic Mesh Generation - Application to Finite Element Methods* Wiley, 1991.

[9] S. Bozek, B. Baccus, V. Senez and Z.Z. Wang, "Mesh Generation for 3D Process Simulation and the Moving Boundary Problem," In Proc. *Simulation of Semiconductor Devices and Processes*, vol. 6, 1995.

[10] S. Solmi, F. Baruffaldi and R. Canteri, "Diffusion of boron in silicon during post-implantation annealing," *J. Appl. Phys.*, 69 (4), pp. 2135-2142, 15 Feb. 1991.

[11] P.M. Fahey, P.B. Griffin and J.D. Plummer, "Point-defects and dopant diffusion in silicon," *Reviews of Modern Physics*, vol. 61, no 2, p.373, April 1989.

[12] R.B. Fair and J.C.C. Tsai, "Quantitative model for diffusion of phosphorus in silicon and emitter dip effect," *J. Electrochem. Soc.*, vol. 124, pp. 1107-1118, 1977.

[13] C.P. Ho, J.D. Plummer, S.E. Hansen and R.W. Dutton, "VLSI process modelling - SUPREM III," *IEEE Trans. Electron Devices*, vol. ED-30, pp. 1438-1453, 1983.

[14] B. Baccus, D. Collard, P. Ferreira, V. Senez and E. Vandenbossche, *IMPACT-4 user's guide*, ISEN, Lille, February 1995.

[15] C. Hill and S.K. Jones, "Modelling diffusion in and from polysilicon layers," *Mat. Res. Soc. Symp. Proc.*, vol. 182, pp. 129-140, 1990.

[16] B. Baccus, D. Collard and E. Dubois, "Adaptive mesh refinement for multilayer process simulation using the finite element method," *IEEE Trans. on Computer aided design*, vol. CAD-11, pp. 396-403, 1992.

[17] C.C. Lin, M.E. Law and R.E. Lowther, "Automatic grid refinement and higher order flux discretizations of diffusion modelling," *IEEE Trans. on Computer aided design*, vol. CAD-12, p. 436, 1993.

[18] J.P. Peng and G.R. Srinivasan, "Non-Linear Visco-Elastic Modeling of Thermal Oxidation of SiO2," *Nasecode VII Tech. Dig*, Ed. J. J. H. Miller, Front Range Press, Boulder, Co., USA, 1991.

[19] P. Conti and W Fichtner, "Automatic grid generation for 3D device simulation," *Simulation of Semiconductor Devices and Processes*, Eds. G. Baccarani and M. Rudan, Vol. 3, pp. 497-505, 1988.

3D Simulation of Topography and Doping Processes at FhG

J. Lorenz[a], E. Bär[a], A. Burenkov[a], W. Henke[b], K. Tietzel [a], and M. Weiß[b]

[a]Fraunhofer-Institut für Integrierte Schaltungen,
Bauelementetechnologie (FhG-IIS-B)
Schottkystrasse 10, 91058 Erlangen, GERMANY
[b]Fraunhofer-Institut für Siliziumtechnologie (FhG-ISiT)
Fraunhoferstrasse 2, 25524 Itzehoe, GERMANY

Abstract

This paper outlines activities carried out at FhG-IIS-B and FhG-ISiT on the development of algorithms and physical models required for the accurate three-dimensional simulation of topography and doping steps in semiconductor technology. The three-dimensional process simulation modules are being developed as parts of the SOLID and the PROMPT process simulation systems.

1. Introduction

More complex device architectures and shrinking device dimensions require major breakthroughs concerning the accuracy of physical models and the functionality of the simulation programs used in the development and optimization of process technologies and devices. On the other hand, progress in computer technology has now made available appropriate computer hardware both in terms of computational speed and main memory to allow for the implementation of advanced physical models in three-dimensional (3D) process simulation tools. Such simulation tools may generally be evaluated and calibrated using two-dimensional (2D) measurements, and then be applied to 3D problems of technological relevance. This is especially important because existing or emerging methods for the characterization of dopant distributions can with sufficient spatial resolution and detection limit only be used for the measurement of 2D structures.

3D effects are important in many of the process steps used for the fabrication of advanced ULSI devices. In Low Pressure Chemical Vapor Deposition (LPCVD), the thickness of layers deposited at the bottom of a contact hole differs drastically from that on the bottom of a trench of the same aspect ratio. In optical lithography, neighboring structures may influence each other via proximity effects. 3D shadowing effects are also important in dry etching steps. Advanced device architectures like LATID (Large Angle Tilted Implant Drain) involve tilted and rotated implants which lead to dopant distributions which depend on the orientation of the implantation beam relativ to the crystal structure and to the mask geometry, and cannot be fully described by a 2D cut, because the standard assumption of a nearly infinite extension into the third orthogonal direction is no more valid. In turn, the appropriate simulation of such structures requires the 3D simulation of both ion implantation and dopant diffusion. Furthermore, corner effects in lateral isolation can only be predicted using the 3D simulation of oxidation.

In addition to the great demands for the development of advanced physical models, the problems highlighted with the examples given above require the development of powerful algorithms e.g. for the calculation of view angles, for the shift of layer surfaces, for efficient calculation of ion implantation into arbitrary non-planar multilayer targets, and for adaptive meshing of dopant concentrations especially in case of moving boundaries. This paper focusses on the activities carried out at FhG-IIS-B and FhG-ISiT on the 3D simulation of topography and doping steps. Work of FhG-IIS-B on meshing in case of moving nonplanar boundaries is only briefly mentioned.

2. 3D simulation of layer deposition

At FhG-IIS-B, a physical model for the simulation of Low Pressure Chemical Vapor Deposition (LPCVD) was developed [1] and implemented into the universal 2D process simulation program STORM [2]. STORM was developed by a consortium of European semiconductor companies, research institutes, and universities within an ESPRIT project. Recently, this model has been rigorously implemented by FhG-IIS-B in three dimensions to allow for an appropriate simulation of layer deposition on 3D structures like contact holes. In the following, the main features of the novel 3D CVD simulation module are described, and some application examples are given.

2.1. Physical LPCVD model

In the physical model implemented for the 3D simulation of LPCVD, the equilibrium reaction rate of reactive particles for each position on the wafer surface is calculated. This model had already been applied for the simulation of LPCVD on structures like long trenches [1] [3] and circular vias [3] where the surface representation is 2D. In our work, a discretization of the surface by a set of triangles allows the application of the model to real 3D cases. In this model, it is assumed that due to thermal equilibrium, the particles have an isotropic velocity distribution far away from the device structure and that their mean free path is far larger than the device feature size, as it is the case for usual LPCVD conditions. Because of the latter, collisions between particles near the wafer surface can be neglected and the particles move along straight lines near the wafer surface. After hitting the surface, a particle can either be incorporated into the deposited layer by reaction, or be emitted again from the surface to escape into free gas space, or hit the surface at another position. Surface diffusion is assumed to play no major role for the redistribution of the reactive particles as it was found for LPCVD of SiO_2, polysilicon, and tungsten [4] [5]. The probability for an impinging particle to undergo a surface reaction is given by the so-called sticking coefficient. As a first order reaction kinetics is assumed, the sticking coefficient is set to a constant for the whole surface not depending on the actual local particle density. The local layer growth rate is the result of a dynamical equilibrium between the incoming particle flux from free gas space or other surface positions and the outgoing flux due to desorption and reaction [1] [3]. Dynamic equilibrium may be assumed because the growth rates which define the time scale for the change of the geometric boundary conditions are very small compared to the particle velocities. Therefore, the time for establishing the equilibrium is very small in comparison to the time interval related to a significant topography change. In consequence, the physical model leads to a large inhomogeneous system of linear equations. The matrix of this

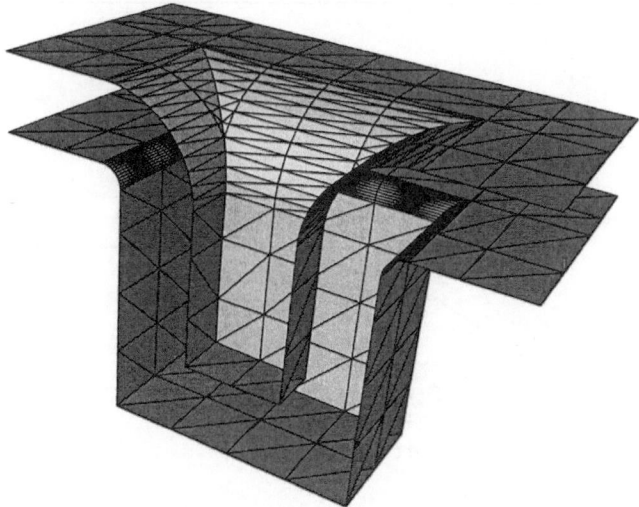

Figure 1: Triangulation of a cubic hole before and after layer deposition

system describes the particle transfer between the individual surface segments and the consumption due to layer growth. The inhomogeneity represents the particle flux from free gas space.

2.2. 3D implementation of the CVD model

As the model requires the accurate knowledge of the surface normal for each position at the wafer surface, we use a segment based approach for the representation of the surface where the wafer geometry is described by a set of N triangles which are shifted during a deposition time step. An example for a triangulation is presented in Fig. 1. The surface of a 1 μm cubic hole is shown before and after deposition of a layer with uniform thickness. For the generation of the initial surface triangulation we used a tool which performs refinement of triangles at convex edges in order to get a final layer profile with round convex edges. The triangulation of the initial structure and of the final surface consists of 640 and 400 triangles, respectively.

For the implementation of the model the particle reaction rate is set to a constant within each triangle. The particle reaction rate is defined as the number of particles which is incorporated per unit time and unit area into the deposited layer by reaction. The particles coming from free gas space and those coming from other surface positions j per unit time and and unit area correspond to the particle fluxes A_i and S_{ij}, respectively. With the sticking coefficient sc, the reaction rate is then given by

$$R_i = sc\,(\,A_i + \sum_{j=1...N, j \neq i} S_{ij}\,) \tag{1}$$

where the index i denotes the triangle under consideration and S_{ij} is the particle flux from triangle j to triangle i. For triangles j which are not visible to triangle i, the corresponding S_{ij} is zero. It is assumed that a desorbing particle has no memory of its incident direction and that the angular distribution of the desorbing particles is given by a cosine law where the particle flux in a specified direction is proportional

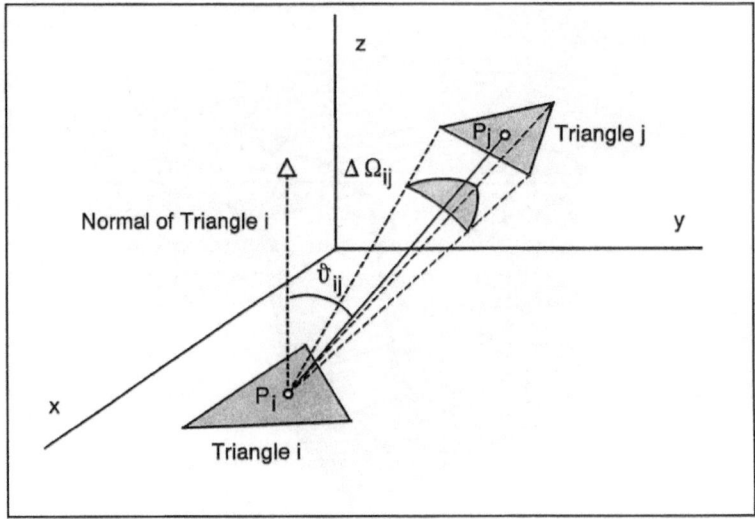

Figure 2: Example for two arbitrary triangles in 3D for which particle transfer has to be calculated by the simulation program

to the cosine of the angle between the surface normal and the direction in question. The reaction rate for each triangle can be calculated by solving a system of linear equations in the unknown reaction rates R_i:

$$\pi R_i - \sum_{j=1...N, j\neq i} T_{ij} R_j = \pi - \frac{1}{1-sc} \sum_{j=1...N, j\neq i} T_{ij} \qquad i = 1...N \qquad (2)$$

where R_i is normalized to 1 at planar sections far from any 2D or 3D structure and T_{ij} is a matrix element describing the probability for a particle desorbing from tringle j to reach triangle i. When deriving this equation, a non-zero growth rate which implies a non-zero sticking coefficient was assumed. The geometry for two arbitrary triangles is shown in Fig. 2. P_i and P_j are the centers of mass of triangle i and triangle j, respectively, ϑ_{ij} is the angle between the normal of triangle i and the vector from P_i to P_j, $\Delta\Omega_{ij}$ is the solid angle covered by triangle j with respect to P_i. For our simulation, the matrix elements T_{ij} are given by

$$T_{ij} = v_{ij} \, (1 - sc) \, cos(\, \vartheta_{ij} \cdot \Delta\Omega_{ij}) \qquad (3)$$

where v_{ij} is the visibility coefficient which is unity if there is direct view between triangle i and triangle j, and zero otherwise. For the calculation of Ω_{ij}, triangle j is projected on the unit sphere around P_i with P_i as projection center and the area of the projected planar triangle is set to $\Delta\Omega_{ij}$. Equation (3) is exact only for infinitesimal small triangles. For a finite triangle size, the matrix elements T_{ij} are given as an integral, which cannot be solved analytically. Therefore, in cases where equation (3) and the procedure for the calculation of $\Delta\Omega_{ij}$ lead to a significant error, the corresponding triangles i and j are each subdivided into smaller triangles. The matrix elements T_{ij} are then obtained by calculating and summing up the matrix elements related to the smaller triangles. Because of the higher efficiency in comparison with direct methods, the solution of the system of linear equations (2) is calculated using an iterative solver. As the result we get the reaction rate for each surface position, i.e. the reaction rate R_i for each triangle. For calculating the shift for each triangle, the 1-D growth rate at planar sections X_{1D} has to be specified as a parameter. As the growth rate is proportional to the reaction rate for each surface position and because we have set $R = 1$ for planar sections, the growth rate for an arbitrary triangle i is given as

$$X_i = X_{1D} \cdot R_i .\tag{4}$$

The update of the surface is done by shifting the triangles according to the calculated growth rates until the desired layer thickness is achieved. For this task, we used a 3D modification of the string-algorithm [6] which calculates the shift for each node depending on the growth rates of the triangles adjacent to this node. The situation for a node with adjacent triangles is shown in Fig. 3a. The vector $\mathbf{n_i}$ denotes the normal of triangle i and γ_i is the angle of triangle i at the node. Using a relative contribution of each triangle which is proportional to γ_i, the algorithm calculates the product of the average normal direction and the average growth rate of the triangles adjacent to the node. In order to avoid artefacts especially at nodes near topography edges or corners the resulting vector is multiplied with a correction factor

$$f_{cor} = \frac{\sum_i \gamma_i}{\sum_i \gamma_i \, cos\varphi_i}\tag{5}$$

to get the final shift $\mathbf{v}$ (defined as distance per time) of the node. Here φ_i denotes the angle between the average normal direction and the normal of triangle i and the sums are taken over all triangles adjacent to the node. While at nearly planar sections of the topography the factor is ≈ 1, for nodes with adjacent triangles having significantly differing normals, f_{cor} is larger than 1 and ensures that the normals of the triangles are hardly changed when shifting the nodes and therefore any artificial surface modification is minimized.

The changes in geometry caused by the deposited layer have to be taken into account by updating the set of reaction rates several times during the simulation. Typically, the simulation of a filling process of a contact hole requires about 10 updates of the reaction rates. Furthermore, the advancement of the surface between two updates of the reaction rates is performed in a number of steps (10 to 100) in order to enable an efficient prevention of surface self intersections. For this purpose, we implemented an algorithm which deletes nearly degenerated triangles of the topography discretization, i. e. triangles with at least one height which is smaller than a given limit. This limit has to be chosen several times larger than the maximum surface displacement during a topography update step to ensure that the triangle is deleted before causing a surface self intersection. The procedure is demonstrated in Fig. 3b, where the two possible cases are shown. Depending on the maximum angle of the triangle in question, it may be necessary to split another triangle into two triangles (case A). In case B, the two nearly degenerated triangles are removed without significantly affecting the remaining triangles. This method has been shown to be a very efficient tool for preventing the formation of surface self intersections.

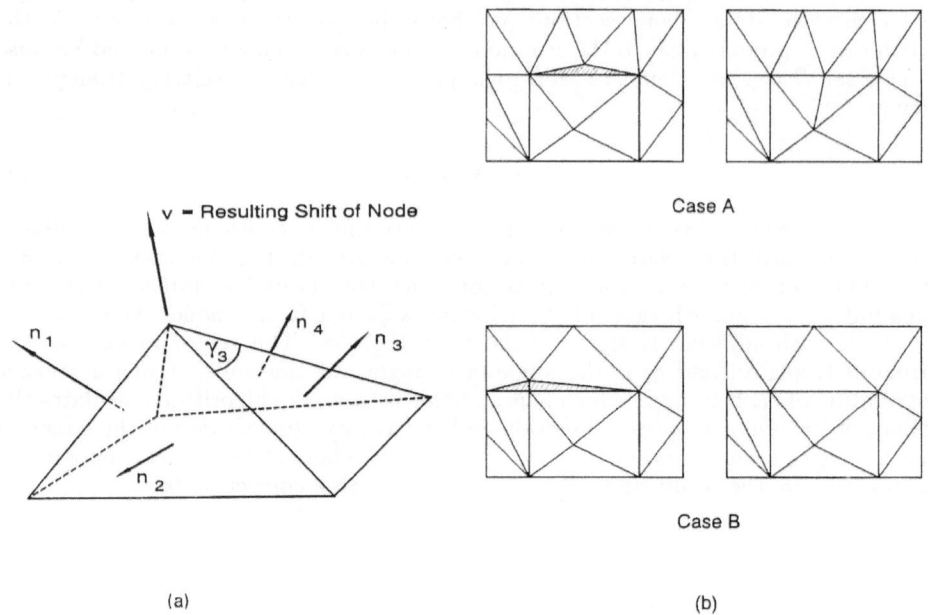

Figure 3: (a) Node with four adjacent triangles contributing to the shift vector; (b) procedure for deleting nearly degenerated triangles (shaded), in two different cases A and B

2.3. Examples for 3D LPCVD simulation

For predicting the profile of a LPCVD layer, it is necessary to know the sticking coefficient and its dependency on the process parameters. For usual LPCVD deposition of low-temperature oxide (LTO), typical values of the sticking coefficient lie in the range between 0.05 and 0.3 [1] [3] depending on the source gases used and on the process parameters. In general, the values for the sticking coefficient were calculated using 2D simulations which were compared to experimental data from geometries like long trenches [1] [3] or using thermodynamics [7]. As the physics is the same in 2D and 3D, these values are also valid for 3D simulation of LPCVD.

As an example, we present the 3D simulation of LPCVD of silicon dioxide from silane and oxygen at 430 ^{0}C using the value sc=0.16 which was found from a 2D deposition experiment [1]. In Fig. 4, two simulations of layer deposition into rectangular holes with an opening of 1 μm · 1 μm are shown. Depositions into a 1 μm deep hole and into a 2 μm deep hole with a deposited layer thickness at planar sections of 0.4 μm are presented. As the solid angle corresponding to the view to free gas space is smaller in case of the deeper hole, the step coverage is significantly reduced for the 2 μm hole compared to the 1 μm hole.

A simulation using 2D calculation of particle transport [1] and a 3D simulation of LPCVD into a 1 μm cubic hole with a sticking coefficient of 0.3 and a deposited thickness at planar sections of 0.4 μm were compared in order to investigate the 3D effect. In Fig. 5, the 2D simulation and the cross sectional view of the 3D simulation are presented. The 3D simulation exhibits a significantly smaller step coverage than the 2D simulation. This is an expected result because the 2D simulation corresponds to an infinitely long trench where for a position at the inner part of the structure the

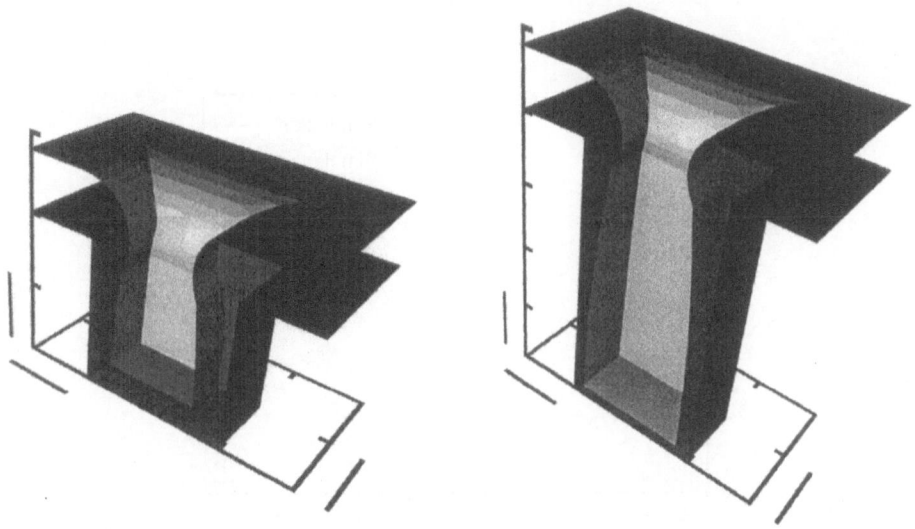

Figure 4: 3D simulations of LPCVD into rectangular holes, assuming a sticking coefficient of 0.16. The bars represent 0.5 μm

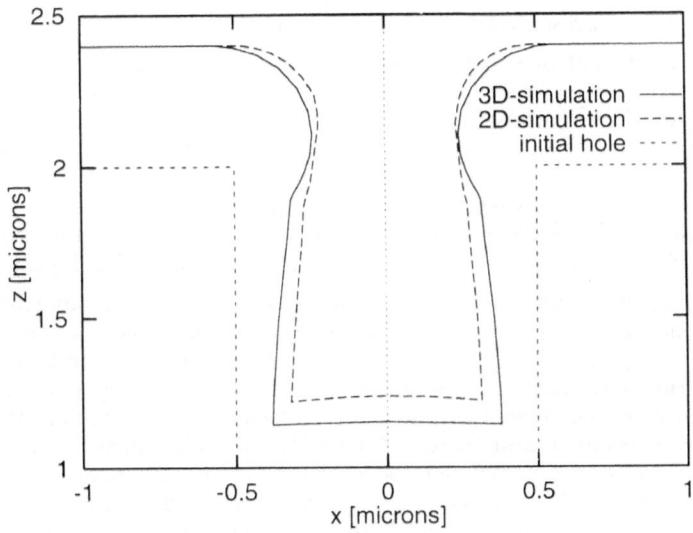

Figure 5: Comparison between 2D and 3D simulation of LPCVD into a 1 μm cubic
hole using a sticking coefficient of 0.3

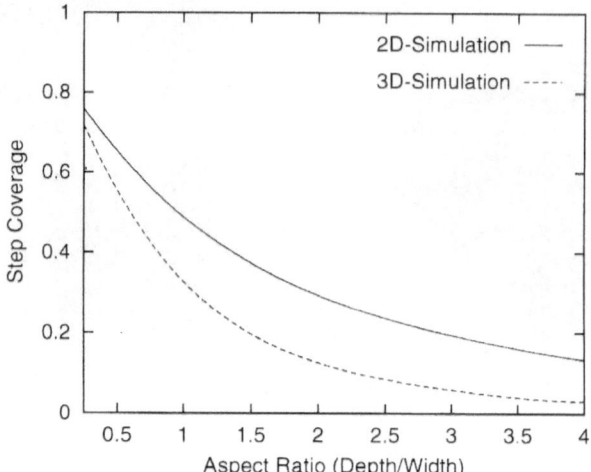

Figure 6: Step coverage of LPCVD layers deposited into rectangular holes with different initial aspect ratios, calculated from 2D and 3D simulations with a sticking coefficient of 0.3

solid angle covering the view to free gas space is larger than for the corresponding position in the hole represented by the 3D simulation.

For a sticking coefficient of sc=0.3 and a deposited thickness at planar sections of 0.2 μm the step coverage (defined as the ratio of minimum thickness to top thickness) was simulated for different initial aspect ratios (height/width) in case of a trench (2D simulation) and in case of a hole (3D simulation). In case of the hole, the cross section perpendicular to the z-axis is a square. The results are presented in Fig. 6. There is a considerable difference between 2D and 3D simulations, especially for large values of the aspect ratio. Therefore, geometries like high aspect ratio contact holes require the use of 3D simulations for accurately predicting the profile of a deposited layer.

In order to compare simulations to experimental data, results of tungsten LPCVD from tungsten hexafluoride (WF_6) and silane (SiH_4) [8] were used. For the simulation of LPCVD in a experimentally prepared contact hole, we approximated the initial hole by a rotational symmetric structure and assumed a sticking coefficient of 0.04, which is the sticking coefficient of the SiH_4 molecules, as a sufficiently high WF_6 partial pressure is assumed and therefore the SiH_4 molecules control growth rate and step coverage. This sticking coefficient is consistent with calculations using the layer growth rate and kinetic theory of gases [7], where a value of 0.05 was predicted for the SiH_4 molecules. The 3D simulation result and a comparison between the cross section of the simulation and experimental data are shown in Fig. 7. The experimental profile was extracted from a SEM cross section of a contact hole with a deposited layer of tungsten. As the experiment agrees very well with the simulation and the sticking coefficient used is consistent with the value predicted from thermodynamics, for this deposition experiment it is admissible to assume one reactive species controlling step coverage.

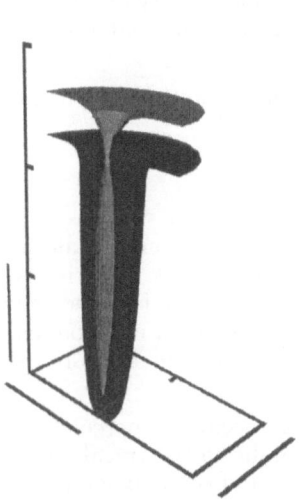
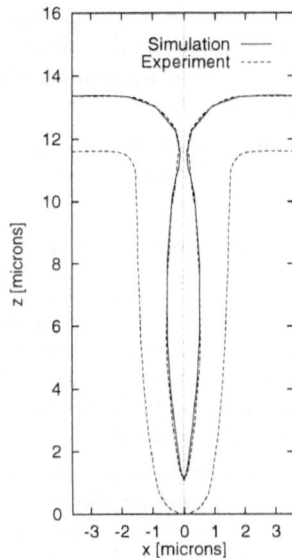

Figure 7: 3D simulation of tungsten LPCVD into a contact hole using a sticking coefficient of 0.04 and comparison of the simulation to a layer profile extracted from a SEM cross section of the contact hole [8]. The bars represent 5 μm

3. Patterning

At the Fraunhofer-Institut für Siliziumtechnologie (ISiT), simulation work is done mainly on the pattern generation processes lithography and dry etching. For optical lithography, the 3D simulation tool SOLID (Simulation of optical LIthography in three Dimensions) [9] has been developed and is now being extended to nonplanar topographies. SOLID features can be grouped into the two major categories optics and resist, covering the following issues:

- Optics
    - partially coherent imaging for arbitrary reticles layouts (conventional and phase-shifting)
    - aberrations of the wafer steppers projection lens
    - spatial filtering techniques
    - various setups of the illumination system (tilted condenser, off-axis illumination)
    - scan and repeat imaging
- Resist
    - positive and negative Novolak and chemically amplified resists
    - propagation into resist with defocussing and bleaching
    - Fickian diffusion of the photoactive compound during post exposure bake
    - 2D and 3D wet development based on various development rate models
    - silylation processes

For the topic of dry etching, the 3D simulator MASTER (Modeling And Simulation of Three dimensional Etching pRocesses) was developed which models dry etching in a plasma by a superposition of a ionic anisotropic and chemical isotropic contribution to the etchrate.

In the following, physical and algorithmic aspects of modeling of wet development and dry etching will be described in more detail.

3.1.   Modeling of wet development and dry etching

Dissolution of exposed photoresist in a development fluid on a non-atomar scale is being considered as a surface limited process. This means that resist material is removed from the actual profile surface which moves with a velocity $\mathbf{v}$. $\mathbf{v}$ is called the development rate and is a function of the local concentration of the resist compounds at the surface and thus the development rate is a (usually smooth) function of the position within the resist layer.

On a more microscopic scale the dissolution is known [10] to occur via percolative diffusion of the alkaline developer molecules along the hydrophilic sites, typically the OH-groups of the phenolic moieties. This gives rise to a gel layer of some nanometers thickness and the percolative nature generates some roughness, whereas our macroscopic model assumes a smooth surface and no intermediate layer.

In a similar way dry etching is modelled as a surface limited process. As in wet development the actual profile moves with a velocity $\mathbf{v}$ normal to the surface. Here $\mathbf{v}$ is called the etchrate and consists of an ionic and a chemical component:

$$\mathbf{v} = (\mathbf{J}_i \cdot \mathbf{n})\mathbf{n} + k_c\mathbf{n} \qquad (6)$$

The vector character of a physical quantity is indicated by bolt face typing. $\mathbf{n}$ is the direction, where the profile is moving, i.e. the inward normal of the actual surface.

$\mathbf{J}_i = k_i\mathbf{n}_i$ takes account of the ionic anisotropic contribution, where k_i is the ionic etchrate and $\mathbf{n}_i$ is the direction of incoming ions, which usually is the negative z-direction. Thus, the scalar product $(\mathbf{J}_i \cdot \mathbf{n})$ contains the cosine between the surface normal and the direction of the ions. The second term $k_c\mathbf{n}$ of eq. (6) describes the chemical component of the etchrate. Note that in the case of wet development this would be the only term in $\mathbf{v}$ with k_c having then the meaning of the position dependent development rate. In the case of dry etching k_i and k_c are constants depending only on the material. Additionally k_i takes the value of zero at those portions of the profile that are not hit by any ions due to shadowing of overlying parts of the profile. Thus, typically $\mathbf{v}$ is a discontinuous stepwise constant function of position in the case of dry etching, wheras it is a smooth function in the case of wet development. See also [11] for a more general formulation of the etchrate including arbitrary angular distributions.

A formal way to describe the evolution of the profile in the time domain is given by the eikonal or Hamilton-Jacobi equation, which will be derived in the following. The profile at etch time or respectively development time t is given implicitly by the "phase" function ϕ as a surface of constant phase or an isophase.

$$t - \phi(\mathbf{r}) = 0 \qquad (7)$$

At an infinitesimal time step later $t + dt$ the point $\mathbf{r} + dt\mathbf{v}$ will obey the above profile equation and one obtains from $t + dt - \phi(\mathbf{r} + dt\mathbf{v}) = 0$:

$$1 - (\mathbf{v} \cdot \nabla\phi) = 0 \qquad (8)$$

Inserting $\mathbf{v}$ from eq. (6) into eq. (8) and using the fact that the inward normal $\mathbf{n}$ of the isophase surface is given by $\mathbf{n} = \nabla\phi/|\nabla\phi|$ one gets the eikonal or Hamilton-Jacobi equation:

$$k_c|\nabla\phi| + k_i(\mathbf{n}_i \cdot \nabla\phi) - 1 = 0 \qquad (9)$$

This first order partial differential equation is of the type of an initial value problem. The initial condition is that ϕ is zero on a given initial surface, which is the profile at the beginning of the etching or developing process. This partial differential equation is then transfered into a system of ordinary differential equations by the so-called method of characterics as described for example by Courant and Hilbert [12]. First a Hamilton function H is derived from the Hamilton-Jacobi equation by substituting $\nabla\phi$ by the momentum $\mathbf{p} = (p_x, p_y, p_z)$:

$$H(\mathbf{r}, \mathbf{p}) = k_c|\mathbf{p}| + k_i(\mathbf{n}_i \cdot \mathbf{p}) - 1 \qquad (10)$$

Here $|\mathbf{p}| = \sqrt{(p_x^2 + p_y^2 + p_z^2)}$ is the modulus of $\mathbf{p}$. The characteristics are the rays in the optical analogon or in the mechanical analogon the trajectories of particles in phase space belonging to the above Hamiltonian and their equations are well known from courses in theoretical mechanics :

$$d\mathbf{r}/dt = \partial H/\partial\mathbf{p} = k_c\mathbf{p}/|\mathbf{p}| + k_i\mathbf{n}_i \qquad (11)$$

$$d\mathbf{p}/dt = -\partial H/\partial\mathbf{r} = -|\mathbf{p}|\nabla k_c - (\mathbf{n}_i \cdot \mathbf{p})\nabla k_i \qquad (12)$$

$$d\phi/dt = \mathbf{p}\partial H/\partial\mathbf{p} = k_c|\mathbf{p}| + k_i(\mathbf{n}_i \cdot \mathbf{p}) \qquad (13)$$

These are the ray tracing equations. The initial conditions are that $\mathbf{r}_0$ must lie on the initial etch surface, $\mathbf{p}_0$ must be normal to the surface , ϕ_0 is zero and $H(\mathbf{r}_0, \mathbf{p}_0) = 0$. H is conserved, as in mechanics it has the meaning of the energy of a particle.

$$H(\mathbf{r}(t), \mathbf{p}(t)) = 0 \qquad (14)$$

Application of energy conservation simplifies the unaccustomed third equation eq. (13) to $d\phi/dt = 1$, which gives $\phi = t$, thus the curve parameter t of the trajectories in eq. (11)-(13) has indeed the meaning of time as in eq. (7). Energy conservation is also useful in handling the discontinuities in k_c or k_i , which would formally result into delta functions on the right hand side of eq. (12). Since $\mathbf{r}(t)$ is continuous and only $\mathbf{p}(t)$ is discontinous, the jump in $\mathbf{p}$ can be calculated from $H(\mathbf{r}, \mathbf{p}_1) = H(\mathbf{r}, \mathbf{p}_2) = 0$. In the isotropic case ($k_i = 0$) the refraction law of Snellius may be derived in this way.

The evolution of the profile is obtained in principle by tracing all rays from the initial profile for the time t, the endpoints of the rays then build up the searched profile. Note that the auxiliary variable $\mathbf{p} = \nabla\phi$ is always normal to the profile, so for nonvanishing anisotropic contributions ($k_i > 0$) the rays $d\mathbf{r}/dt$ eq. (11) are usually not normal to the profile. A second note applies for dry etching, where k_c and k_i are piecewise constant, which gives straight lines for the rays as long as no discontinuities occur.

3.2. Numerical methods

There a several numerical methods to simulate profile evolution:

- ray tracing methods

- segment or surface advancement methods

- advection and diffusion methods

- cell-removal methods

Methods based on ray tracing take advantage of the fact that the ordinary differential equations (eqs. 11, 12) can be solved very fast and accurately and for each ray independently. The complicated numerical problem consists in reconstructing the profile from the end points of the rays. The important point to notice is that rays can (and typically do) cross and then propagate into regions where no material is present anymore. Thus certain parts of the calculated profile, so-called loops, are fictitious and have to be discarded by some clever algorithm. This can be a local problem in the sense that rays starting from nearby positions cross, which in the optical analogon would lead to caustics. More serious to remedy is the global case, when holes or bridges or more complex topological structures may be formed. This approach was implemented for example in an older version of ISiT's X-ray lithography simulator XMAS [13].

In the simulation of dry etching the right hand side of the differential equations (eqs. 11-13) might not even be known apriori due to shadowing effects. The flux and thus the etch rate rather depends on the evolution of the etch profile, so for each intermediate time step the profile has to be constructed and the shadowing effect has to be newly calculated.

For these kinds of problems segment based methods seem to be more appropriate. This means that the actual profile is discretized by a (usually triangular) mesh and the nodes are moved in small timesteps according to the ray tracing equations (eqs. 11-13) or using directly eq. 6 . The local loop problem can then be solved by proper mesh maintenance [14], the global problem of selfintersections remains to be intricate. For a detailed discussion see the thesis of E. W. Scheckler [15].

Methods that work on a fixed 3D mesh are the advection, the diffusion and cell-removal methods. In the advection method the Hamilton-Jacobi equation (eq. 9) is directly tackled [16] [17]. The phasefunction ϕ is given an initial definition ϕ_0 over the whole 3D simulation region, which has to be compatible with the initial surface, and then in each time step the phase ϕ is updated in all cells. Local and global loops can be easily avoided, since already removed cells are recognized by their value of ϕ. The etched profile is given implicitly by $\phi(\mathbf{r}) = t$ and is constructed by using a standard isosurface algorithm. In a similar way a 3D diffusion-reaction equation instead of the Hamilton-Jacobi equation is sometimes taken [18]. This can be thought of as a different physical model that takes into account the formation of a gel layer in wet development [10], or alternatively, it is taken just as an approximation of the nasty Hamilton-Jacobi equation by the well understood fickian diffusion equation. The main drawback of these two methods is that in each time step the phase or concentration has to be updated in the whole bulk domain and not only on the surface, so CPU time is typically huge as compared to the segment or ray tracing methods, which are surface limited. Furthermore, the convergence seems to be poor [16], when the computing mesh is refined.

Therefore the old cell-removal method [19] [20] was chosen in SOLID and MASTER as the algorithm for surface evolution. Our improved version of the algorithm [21] will be described in more detail now. In describing it we will adopt the etching jargon.

The 3D region under consideration is divided into rectangular, elementary cells. An integer valued counter representing the cell volume is assigned to each cell, which is decremented during the profile evolution. It is assumed that the etch rate is constant throughout each cell. A cell is removed when its counter is zero. Various cell types are defined depending on the number of sides exposed to the etching plasma. In Fig. 8 cells are displayed, having one, two and three sides open for the etching attack.

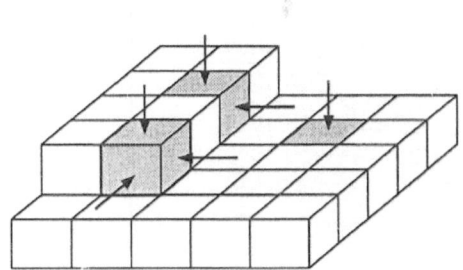

Figure 8: Different cell types are defined depending on the number of sides exposed to the plasma

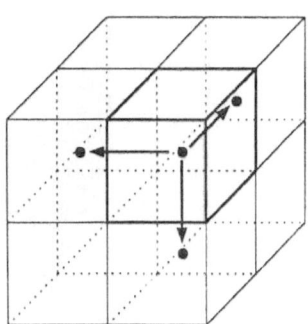

Figure 9: Basic scheme of the overkill concept: spillover of decrement is distributed to adjacent cells

The total etch time is divided into single time steps and all surface cells are treated once during each time step. For cells with one surface exposed to the plasma, the amount subtracted from the cell counter, i.e. the counter decrement, is determined from the etch rate, the cell volume and the size of the exposed surface area of the cell. The decrement is modified when several faces of the same cell are exposed to the plasma. In order to do this, an effective surface area of the cell is defined, which essentially is the cell's surface area projected onto to the global etch profile. More precise determination of this area and of the surface normal may be obtained by taking into account also the counters of neighbor cells. Furthermore, the cells with an open unshadowed top side are given an additional contribution to include the anisotropic etch component.

If, during the course of an etch simulation, a decrement exceeds the counter value, which is currently assigned to a particular cell, a hangover or spillover of decrement results, which we call overkill. If for example, the counter of a given cell is 10 and the decrement is 30 an overkill of 20 results. After each time step the resulting overkill of removed cells is distributed to adjacent cells in the same direction the etch attack has previously taken place. Fig. 9 shows one cell, which is to be removed (dark frame), and the occurring overkill is distributed to the three neighboring cells. The same approach is taken for other cell types, e.g. cells with two, four and five sides exposed to the plasma. After each time step the set of surface cells must be updated as developed cells are removed, and new cells must be admitted.

If the overkill is distributed to a cell, which already has a very small counter, a secondary overkill may result, in this case the overkill again is distributed to neighboring cells according to the scheme displayed in Fig. 9. Three major advantages can be

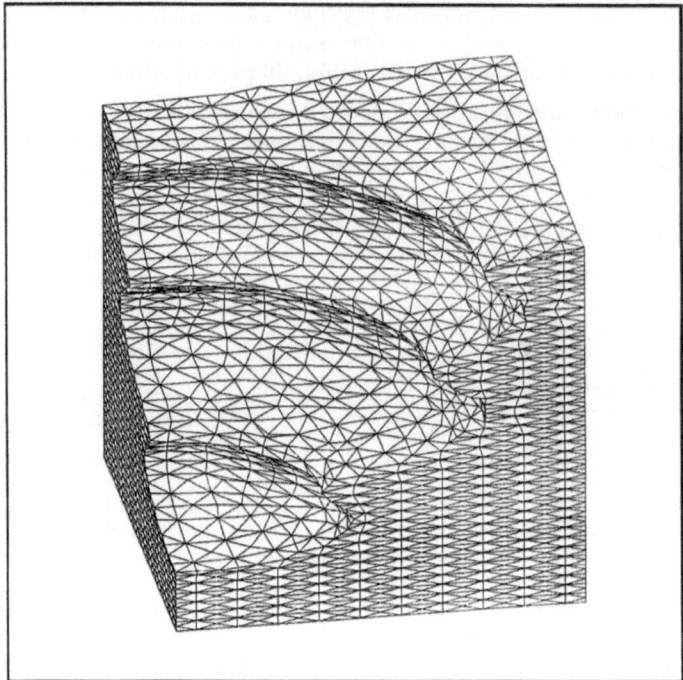

Figure 10: Surface elements after triangulation

attributed to the overkill concept: first the computing time required is reduced, since larger time steps are allowed. Second, any loss of cell counter, which can be translated into a reduction of etch rate, due to the applied time discretisation is avoided, and accuracy of the method is retained. Third, certain anisotropies that are inherent to the pure cell-removal without overkill are avoided. The overkill concept is of essential importance for our implementation of the cell-removal algorithm.

At the end of the last time step all remaining surface cells represent the resulting profile. In a final smoothing operation the center coordinates of the cells are corrected according to their counter values. The counter corrected positions of the neighboring cells can optionally be taken into account. The corrected cell centers are then used for a triangulation, so that the eventually evolving three-dimensional, apparently smooth surface consists of a set of triangular surface elements. As an illustration Fig. 10 shows one quarter of a resist contact hole, which is not completely developed. For displaying the etched or developed profile a slightly modified version of the hidden surface algorithm by Newel, Newel, and Sancha [22] has been adopted. For an enhanced visual presentation the surface elements are shaded according to their orientation with respect to the viewer.

The main advantages of the cell-removal algorithm, if compared with the mentioned alternative algorithms, are the speed combined with numerical stability. The cell-removal algorithm never falls into an unstable state, because cells are removed one by one, so that no loops can occur, if this method is used. Moreover, the convergence of the cell-removal method is superior, since in our experience no change in the resulting profile can be noticed, if the underlying mesh is refined, as is the case for segment and advection algorithms.

3.3. Application

This sections presents some examples for applications of topography evolution algorithms to lithography and dry etching. To fully examine the scaling of lithographic features with new materials, exposure tools and topography features, the 3D nature of resist dissolution and pattern transfer must be considered. Particularly in lithography 3D issues, e.g. defects or optical proximity effects, are becoming bottle necks for implementation of more and more advanced technologies. Proximity effects are even enhanced by the use of sophisticated exposure techniques like off-axis illumination or phase-shifting masks.

A first example demonstrates the extreme stability and robustness of the cell-removal algorithm. As a test vehicle reticle defects are being considered here. A reticle with programmed defects on it was printed on a g-line, 0.55 NA Canon stepper with $\sigma = 0.5$. The resist was the TSMR V3 system of Tokyo Ohka, spun on a bare silicon substrate with final thickness of 1.5 micron. The size and position of a sample defect centered between two 0.65 micron lines is depicted in Fig. 11. Note that all numbers are given in wafer scale dimensions. Fig. 11 displays the SEM of the developed resist profile and a 3D view of the profile after development simulation. A cut along line A-B through the simulated profile shown in Fig. 12 reveals that the developer has actually broken through the resist bridge at positions where the anti-nodes of the standing wave interference patterns were located. This demonstrates that highly complex topographical structures are correctly handled by this type of algorithm.

An example for an entire process sequence of a metal level (or a poly-silicon gate level) pattern process is given in Fig. 13 to Fig. 16.

Fig. 13 shows the profile of a developed resist layer after imaging two vias or contacts onto it and after a subsequent dry etching step in which the via hole was transfered into the underlying (oxide) layer. The left part of Fig. 14 shows the profile after resist stripping. In a deposition step the entire profile is covered with, for instance, a metal layer (right part of Fig. 14). Note that deposition is treated here in a very simplified view as a conformal deposition process. After deposition another lithography step and dry etching step are performed for the patterning of the metal layer, see Fig. 15. Finally, the entire profile (contacts/vias plus wiring level) after resist stripping, and the free standing metal bars after an additional oxide removal step (lift off), which could be useful for micromechanics application, are shown in Fig. 16.

For the entire sequence this type of simulation requires roughly 15 minutes on a current state of the art workstation. It should be noted though, that the selected number of meshpoints does have a marked impact on the resolution and thus on the computing resources required.

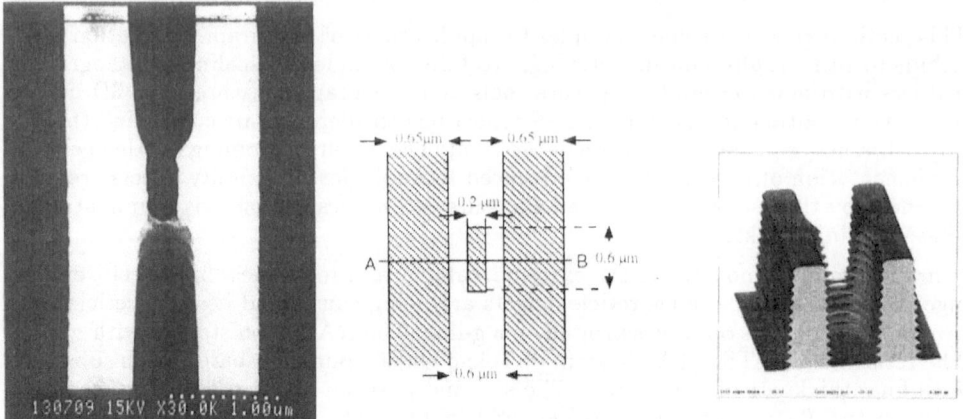

Figure 11: 0.2 μm · 0.6 μm rectangular defect centered between two 0.65 μm lines: SEM on the left, simulated resist profile on the right, 0.55 NA, g-line exposure

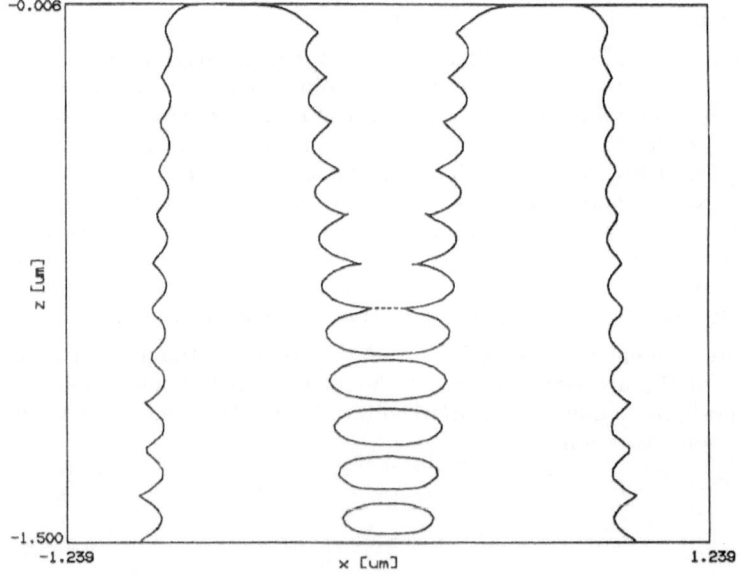

Figure 12: Cross section along the line A-B through simulated profile of Fig. 11

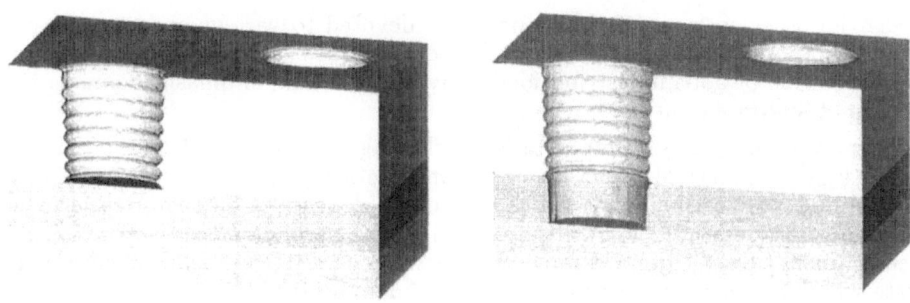

Figure 13: Left: profile of developed resist layer after imaging two via holes; right: via holes after dry etching

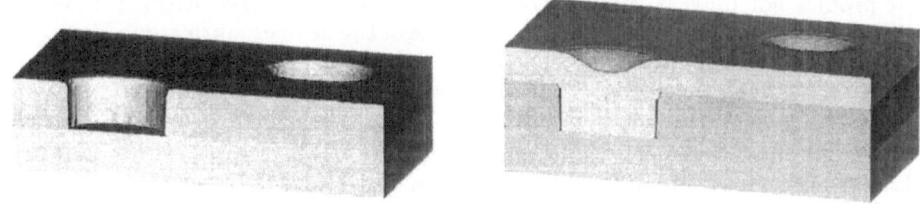

Figure 14: Left: etched vias after stripping; right: conformal deposition of metal layer

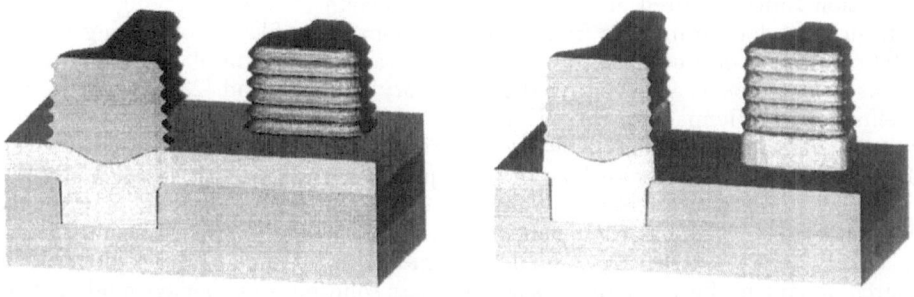

Figure 15: Left: profile of developed resist after patterning two lines with via pads; right: Metal lines after dry etching

Figure 16: Left: metal lines after resist stripping; right: free standing metal lines after oxide removal

4. 3D simulation of doping processes

Whereas the other sections of this paper were devoted to various aspects of topography simulation, in this chapter an outline of activities carried out at FhG-IIS-B on the simulation of dopant distributions is given. The main emphasis is put on the simulation of ion implantation.

3D effects are crucial for ion implantation technology and its simulation in various respects: First, the crystal structure of silicon drastically influences the implanted dopant profiles due to channeling which is strongly dependent on the orientation of the implantation beam relative to the crystal axes. Second, in case of tilted and rotated implantations the ion beam is no more within the 2D vertical cut parallel to the gate which is general used as the simulation domain for 2D tools. Additionally, nonideal mask edges as well as tilted implants require the use of appropriate multilayer models to describe the influence of the different stopping powers of layers which are penetrated by the ion beams. Third, in advanced ULSI devices both the channel length and the channel width are well below 0.5 μm which means that details of the dopant profiles not only at mask corners but also at mask edges strongly influence both the electrical characteristics and the life time of the devices fabricated. Moreover, the shallow junctions necessary for such devices require a reduced thermal budget. In consequence, many details of the implanted dopant profiles are no longer destroyed by subsequent annealing steps.

The two main approaches for the simulation of ion implantation are Monte-Carlo simulations and the use of analytical expressions. Both have their benefits and drawbacks: In Monte-Carlo simulations basic physical models are being used which lead to a large domain of applicability in terms of the process parameters and the geometries to be addressed. The main drawbacks of this approach are the very high computation times required and the difficulties encountered when trying to extract the physical model parameters from measured dopant profiles. Concerning the 2D and 3D simulation of ion implantation, a considerable reduction of the computation time has recently been achieved by the trajectory split method [23] developed in a joint effort of TU Vienna and FhG-IIS-B. The 3D simulation of ion implantation into a typical device structure requires, however, still computation times of at least several hours on a standard workstation. In consequence, a considerably faster method is required for standard applications. This need can be fulfilled by the use of advanced analytical models. Whereas their benefits are the considerably smaller computation times and the easier extraction of model parameters from one- and two-dimensional measured dopant profiles, the drawback is the less solid physical background of these models and, in turn, the need to perform more comparisons with experiment for model calibration.

In the following, results obtained at FhG-IIS-B on the simulation of ion implantation using analytical models are summarized, referring to the three areas mentioned above.

4.1. 3D analytical models for the simulation of ion implantation

Within the last couple of years advanced analytical models have been developed at FhG-IIS-B for the simulation of ion implantation into amorphous targets. These include the use of Pearson distributions both for the vertical and the lateral dopant profile in the convolution integrals to be calculated [24], multilayer models which take into account the influence of different stopping powers of layers on the vertical dopant profile [25] [26] and on the lateral dopant profile [27], and the depth-dependence of the lateral dopant distribution [24] [2] [28]. Because of the great problem to sufficiently characterize lateral dopant distributions by 2D measurements the use of the

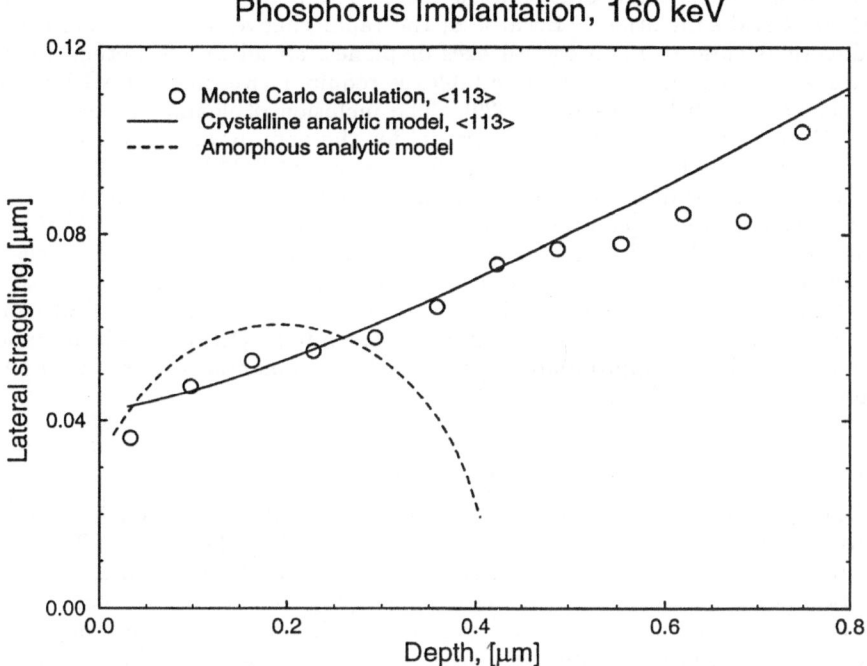

Figure 17: Comparison between Monte-Carlo results and analytical models for the lateral range straggling versus depth for the implantation of phosphorus into crystalline silicon at an energy of 160 keV

Boltzmann transport program RAMM [24] for the efficient calculation of range parameters in all kinds of amorphous targets has been very important: With RAMM a full table of vertical, lateral, and mixed range moments for all energies between 1 keV and some MeV can be calculated for an arbitrary combination of implanted ion and amorphous substrate material within a computation time of about one hour on a standard workstation, and used in the database for subsequent simulations.

For the simulation of ion implantation into crystalline targets, however, the required range parameters cannot be obtained from RAMM, and models must be enhanced to take into account residual channeling. In Fig. 17, the lateral range straggling versus depth is shown for the implantation of phosphorus into crystalline silicon at an energy of 160 keV. The analytical model for amorphous targets is compared with crystalline Monte-Carlo simulations using the Monte-Carlo module of the VISTA program from TU Vienna [23]. The figure shows that the amorphous model drastically underestimates the lateral range straggling towards the tail of the implanted dopant distribution. By using range parameters extracted from the Monte-Carlo simulation, however, the parabolic-exponential approach used in the analytical model for the depth-dependent lateral spread can also be applied to crystalline targets and yields good agreement with the Monte-Carlo simulation, as shown in Fig. 17. In consequence, the analytical approach can be extended to crystalline targets by replacing the table of range moments calculated with RAMM by a table which takes into account the channeling in crystalline material. In this table the range parameters depend not only on the implantation energy, but also on the kind of crystalline material and on the tilt and rotation angles. The parameters required are extracted from Monte-Carlo simulations. Because of the rapid change of parameters for ion beam directions close to main crystal axes or planes, an advanced method for the efficient adaptive organization of these tables is required. Results of FhG-IIS-B on this subject are published elsewhere [29]. The inclusion of channeling effects into 2D and 3D ion implantation models is crucial for the appropriate simulation of advanced devices. This has been shown e.g. for the case of FOND (Fully Overlapped Nitride Mask Defined) devices where details of the polysilicon grain structure drastically influence both the dopant distributions in the silicon and the electrical behavior of the device [30] and for LATID (Large Angle Tilted Implant Drain), as discussed below. In consequence, channeling both in vertical and in lateral direction must be taken into account in the 3D simulation of ion implantation. The analytical model for channeling outlined above has been implemented into STORM. This is an important prerequisite for the appropriate simulation of ion implantation steps during the fabrication of advanced devices.

4.2. 3D algorithms for the simulation of ion implantation

The effect of wafer tilt and rotation is presently not fully considered in available simulators which use analytical models. Results published by the group of Tasch [31] are limited to small tilt angles and furthermore restricted by the use of Gaussian distributions with depth-independent range straggling in lateral direction. In the new STORM ion implantation module, Pearson distributions with depth-dependent parameters are used in lateral direction to describe the implantation into crystalline silicon. Furthermore, the module has been extended to correctly calculate a 2D cut through a 3D implantation profile: In case of nonzero tilt and rotation angles, the new module first calculates the moments of the 3D point response function, using the advanced models described above. Then, a novel algorithm is applied which calculates the moments of the 2D projection of the 3D point response function to the simulation plane specified. In addition to ion energy and dose, now tilt and rotation angles are considered in the module. No approximations are involved in the projection algorithm. In consequence, ion implantation into a 2D structure, e.g. a long mask window, can be appropriately simulated also in cases where the implantation beam is outside the simulation plane: The module is capable to deal with the 3D physics of implantation into a 2D structure.

In Fig. 18, the influence of the rotation angle on the implanted dopant distributions is demonstrated for the example of a boron implantation with an energy of 5 keV and a tilt angle of 45^0. Crystalline parameters were used in the simulation. The symmetry between the results for rotation angles of 60^0 and 120^0 serves as a test of the implementation. The larger depth of the equiconcentration lines for 90^0 rotation are due to the Gaussian approximation used in lateral direction which fixes the value of the lateral kurosis to three. Applying Pearson distributions in lateral direction allows for the use of the correct value of the kurtosis resulting from the projection algorithm outlined above, and removes the difference in the penetration depth at different rotation angles shown in Fig. 18.

Fig. 19 shows the donor distributions in a LATID MOSFET at the end of device processing, simulated under different assumptions about implantation models and about implantation conditions at the stage of the LATID implantation. The lightly doped source/drain region of the device was formed by a double ion implantation of phosphorus with an energy 20 keV and a dose of $3 \cdot 10^{13} cm^{-2}$ under a tilt angle of $\pm 45^0$ and different rotation angles. The simulation plane chosen is parallel to a (100)-type crystallographic plane of the silicon oriented normal to the wafer flat.

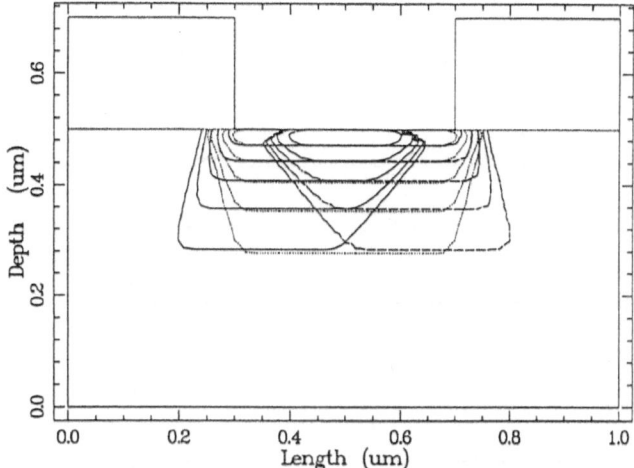

Figure 18: Distributions of boron ions implanted at an energy of 5 keV, dose $5 \cdot 10^{14} cm^{-2}$, and $45°$ tilt for different rotation angles. Solid line: $60°$; dotted line: $90°$; dahed line: $120°$. Lines denote concentrations of 10^{15}, 10^{16}, 10^{17}, 10^{18}, and $10^{19} cm^{-2}$, respectively. Gaussian approximation used for the lateral distributions

The default ion implantation model of STORM-5.0 does not take into account the channeling dependence on the ion beam orientation relative to crystallographic directions, therefore this model predicts a very small change in the dopant distribution in dependence of the rotation angle. Another reason for the small effect of rotation when simulated with the default model is an approximately equally large values of the lateral and projected range stragglings at this energy of phosphorus ions. The phosphorus distribution at larger rotation angles, using the default model, is slightly shallower and penetrates not so much under the gate edge, but the changes are too small to be distinguished well on the picture. Therefore, to keep a clear picture, only the result calculated with the default model for a rotation angle of $0°$ is presented.

In the crystalline based model, the range parameters depend on the beam orientation relative to the crystal lattice, therefore for each rotation angle a different set of the range parameters was used. Furthermore, the projected range straggling is due to channeling usually significantly larger than the lateral straggling in the crystalline mode, especially when the impact direction is close to one of the main crystallographic directions. In this example the rotation angle of $45°$ corresponds to the (110) crystallographic axis of silicon, and a rotation angle of $36°$ is close to this direction as well. Therefore, the crystalline based model predicts a much deeper penetration of the ions when implanted at these angles. Generally, the shape of the distribution depends on the orientation of the ion beam due to the physical reason of channeling and due to the geometrical reason of the angles involved.

The approach mentioned above is sufficient for situations where the device geometry can be considered as 2D and where implantation directions are either within or outside the 2D simulation domain defined by the geometry. In case of advanced ULSI devices,

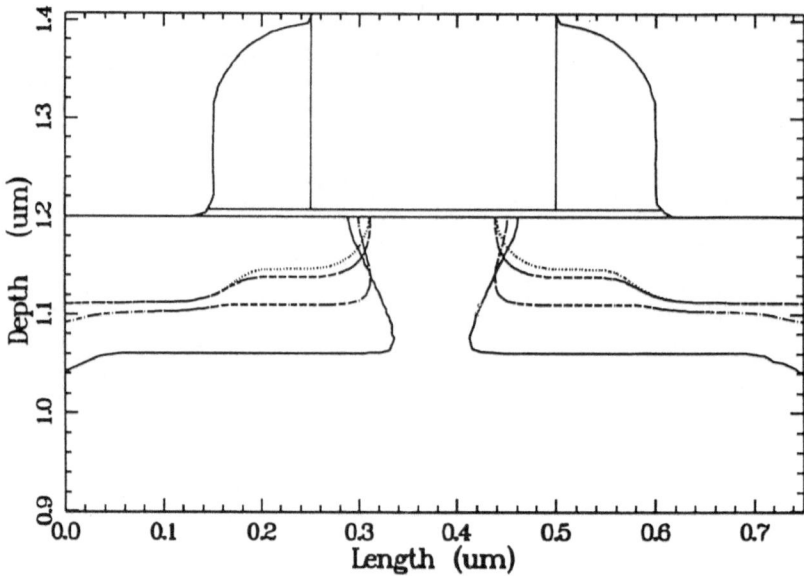

Figure 19: Simulated equiconcentration lines of donor distributions at the level of $5 \cdot 10^{17} cm^{-2}$ in a LATID device using different model assumptions. Dotted lines: Default STORM 5.0 model (assuming an amorphous target) for $\pm 45^{0}$ tilt and 0^{0} rotation; other lines: analytical model for crystalline targets for $\pm 0^{0}$ tilt and 0^{0} (dashed), 36^{0} (dashed-dotted), and 45^{0} rotation

however, the real 3D geometry must be considered, because both the channel length and the channel width are comparable to the lateral extension of the implanted dopant distribution. This requires the development of a 3D algorithm for the simulation of ion implantation, which includes the advanced analytical models described above. In the development of this algorithm it must be considered that the implanted dopant distribution at each meshpoint is calculated from a 2D convolution integral, using vertical distributions which depend on the thickness of the material on top of the point in question via the multilayer model used [25] [26], and lateral distributions which depend on this thickness via the depth-dependent range parameters. Therefore it is very important to efficiently calculate for each convolution point the thickness of the material which is above the point in question, in direction of the implantation beam. In the following, this value is called the effective thickness. In case of multilayer structures, it is the sum of the thicknesses of each layer, scaled by the projected range in that layer. Because the wafer geometry may change drastically and even discontinuously in lateral direction, a straightforward calculation of the thicknesses required for each convolution point would involve the calculation of the intersections between the implantation beam which passes through the point in question and all surface segments. In a typical example, about 10 000 convolution points are needed for each of the about 10 000 mesh points. Furthermore, it is assumed that the device geometry is discretized by about 1 000 triangles. Simple arithmetics shows that the computation time needed on a standard workstation would be about one day. Any drastic speedup of this standard approach would require elaborate search procedures for the intersections with the surface triangles and an adaptive meshing for the convolution integral, depending on the smoothness of the surfaces.

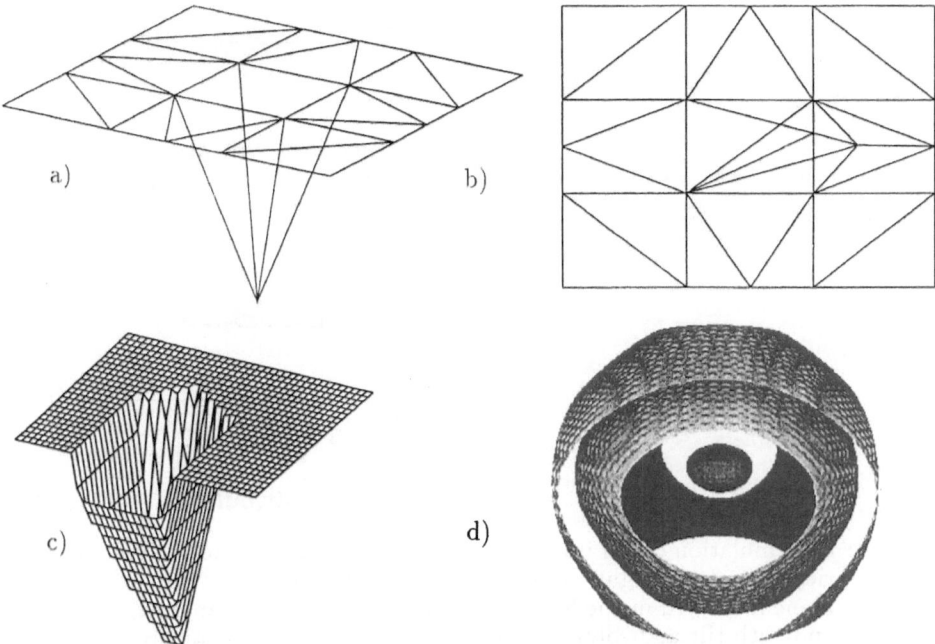

Figure 20: Principle of the novel algorithm for the simulation of ion implantation using analytical models: a) initial triangulation for a hole; b) triangulation of projection parallel to the implantation beam onto a plane; c) effective thickness for triangulation of b); d) 3D point response function for implantation of boron into silicon with an energy of 100 keV

In order to solve this problem a novel algorithm has been developed at FhG-IIS-B which divides the simulation of ion implantation into a preprocessing step in which all relevant information is extracted from the device geometry, and the calculation of implanted dopant distributions from this intermediate data. A 2D version of this algorithm was outlined earlier [32]. In the first step, all surface triangles which are above the mesh plane under consideration are projected down to that mesh plane, parallel to the implantation beam, and subsequently inserted into a triangulation of that plane. This involves the subsequent refinement of all triangles which were projected to the plane before and are intersected by the newly projected triangle. At each node of the projected triangles, the thickness of the material on top of the point is calculated in parallel to the projection algorithm. The sign of the scalar product between the original surface triangle and the implantation beam is used to make sure that overhanging structures are treated correctly. The result of this algorithm is a triangulation of an arbitrary horizontal plane, for which the exact value of the effective thickness is known at each node, without any discretization error compared with a direct calculation. The main advantage is now that the effective thickness behaves linear within each of the projected triangles, which means that the values needed at the convolution points can be calculated by bilinear interpolation instead of calculation of all intersections with surface triangles. In consequence, the computation time drops by some orders of magnitude especially for examples with large numbers of mesh points and surface triangles. 3D examples of moderate complexity calculated with the current version of the algorithm which is not yet fully optimized needed computation times of a few minutes.

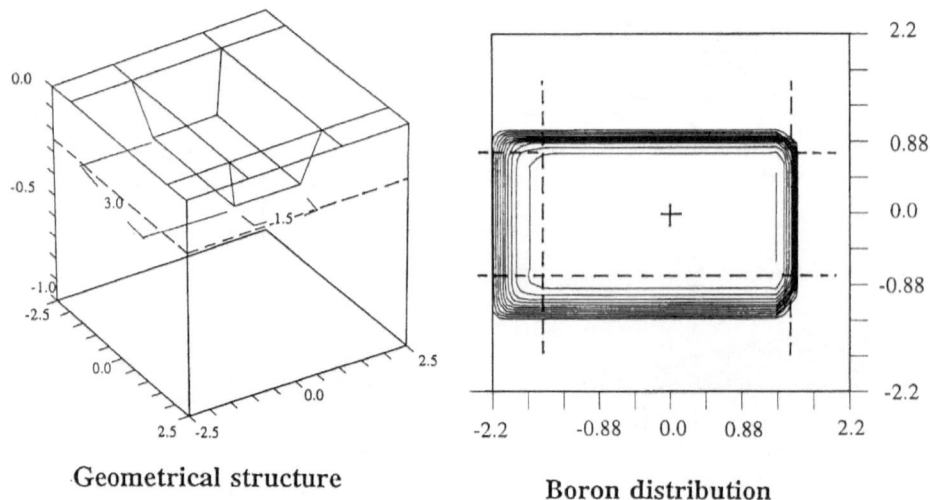

Geometrical structure   Boron distribution

Figure 21: 3D simulation of ion implantation through a mask window. Left: geometrical structure; right: equiconcentration lines of boron in the horizontal
plane indicated in the left figure. Boron implanted at an energy of 25 keV
and with tilt and rotation angles of 45^0 and 30^0, respectively

In Fig. 20, the basic principle of the novel 3D algorithm is outlined. In Fig. 21 the
equiconcentration lines of boron in a horizontal plane after implantation with nonzero
tilt and rotation angles, simulated with the novel algorithm, are shown. Due to tilt
and rotation the dopant profile is shifted with respect to the mask window.

4.3. Meshing aspects

A crucial problem for the 3D simulation is the generation and adaptation of a numerical mesh which sufficiently resolves the dopant distributions with an acceptable
number of mesh nodes. The solution of this problem is especially difficult for nonplanar moving interfaces where the Cauchy boundary conditions which describes the
dopant flux between layers must be accurately discretized in order to assure that the
profile and the total concentration of the dopant in each layer are correctly reproduced.

During an ion implantation step, the dopant distribution in a device in general changes
more rapidly than during an individual time step within the simulation of diffusion
or oxidation, especially if the implantation step is approximated as an instantaneous
event. This makes mesh adaptation during ion implantation mandatory. At FhG-
IIS-B, the effective thicknesses discussed above are being used for the adaptation of
an Octree-based mesh during ion implantation. A special meshing algorithm for the
error control in the discretization of Cauchy boundary conditions is being developed
and will be published separately. The activities on software development are linked
with those on the development of physical models, for which some results are been
presented at SISDEP '95 [23] [29] [33] and ESSDERC '95, [30] [34], respectively.

5. Conclusions

Algorithms developed at FhG for the 3D simulation of various process steps have been outlined together with model results which have proven to be especially important for 3D simulation. The algorithms and models developed have been implemented into the multidimensional PROMPT process simulation system which is being developed by a consortium of European semiconductor companies, software houses, research institues, and universities.

Acknowledgements

Part of this work has been carried out within the JESSI projects BT1/BT11 ADE-QUAT and BT8B PROMPT, funded by the European Union as ESPRIT projects 7236/8002 and 8150, respectively. The authors would like to acknowledge important contributions from S. List and J. Pelka.

References

[1] H. Wille, E. Burte, H. Ryssel, "Simulation of the Step Coverage for Chemical Vapor Deposited Silicon Dioxide", *J. Appl. Phys.*, vol. 71, pp. 3532, 1994.

[2] J. Lorenz, C. Hill, H. Jaouen, C. Lombardi, C. Lyden, K. de Meyer, J. Pelka, A. Poncet, M. Rudan, S. Solmi, "The STORM Technology CAD System", in: Technology CAD Systems (eds. F. Fasching, S. Halama, S. Selberherr), pp. 163, Springer Verlag, Wien, 1993.

[3] M. M. IslamRaja, M. A. Cappelli, J. P. McVittie, K. C. Saraswat, "A 3-Dimensional Model for Low-Pressure Chemical Vapor Deposition Step Coverage in Trenches and Circular Vias", *J. Appl. Phys.* vol. 70, pp. 7137, 1991.

[4] L. Y. Cheng, J. P. McVittie, K. C. Saraswat, "New Test Structure to Identify Step Coverage Mechanisms in Chemical Vapor Deposition of Silicon Dioxide", *Appl. Phys. Lett.*, vol. 58, pp. 2147, 1991.

[5] J. P. McVittie, J. C. Rey, L. Y. Cheng, M. M. IslamRaja, K. C. Saraswat, "LPCVD Profile Simulation Using a Re-Emission Model", Proc. IEDM 90, pp. 917, 1990.

[6] R. Jewett, "A String Model Etching Algorithm", SAMPLE Report No. SAMD-3, University of California, Berkeley, 1979.

[7] J. E. J. Schmitz, R. C. Ellwanger, A. J. M. van Dijk, "Characterization of Process Parameters for Blanket Tungsten Contact Fill", in: Tungsten and Other Refractory Metals for VLSI Applications III (ed. V. A. Wells), pp. 55, MRS Pub., Pittsburgh, 1988

[8] A. Hasper, J. Holleman, J. Middelhoek, C. R. Kleijn, C. J. Hoogendoorn, "Modeling and Optimization of the Step Coverage of Tungsten LPCVD in Trenches and Contact Holes", *J. Electrochem. Soc.*, vol. 138, pp. 1728, 1991

[9] W. Henke, G. Czech, "Simulation of Lithographic Images and Resist Profiles", *Microelectronic Engineering*, vol. 11, pp. 629, 1990

[10] T. F. Yeh, A. Reiser, R. R. Dammel, G. Pawlowski, H. Roeschert, "A Scaling Law for the Dissolution of Phenolic Resins in Aqueous Base", *SPIE* vol. 1925, pp. 570, 1993

[11] V. K. Singh, E. H. Shaqfeh, J. P. McVittie, "Simulation of Profile Evolution in Silicon Reactive Ion Etching with Re-emission and Surface Diffusion", *J. Vac. Sci. Technol. B*, vol. 10, no. 3, pp. 1091, 1992

[12] R. Courant and D. Hilbert, "Methods of Mathematical Physics" (Wiley, New York, 1974) Vol. II , page 62

[13] J. Chlebek, H.-L. Huber, H. Oertel, M. Weiss, R. Dammel, J. Lingnau, J. Theis, "Computer-aided Resist Modelling with Extended XMAS in X-ray Lithography", *Microelectronic Engineering*, vol. 9, pp. 629, 1989

[14] J. J. Helmsen, A. R. Neureuther, "3D Lithography Cases for Exploring Technology Solutions and Benchmarking Simulators", *SPIE*, vol. 1927, pp. 383, 1993

[15] E. W. Scheckler, Ph. D. Dissertation, University of California, Berkeley, Nov. 1991

[16] J. J. Helmsen, M. Yeung, D. Lee, A. R. Neureuther, "SAMPLE-3D Benchmarks Including High NA and Thin Film Effects", *SPIE*, vol. 2197, pp. 478, 1994

[17] M. Komatsu, "Three Dimensional Resist Profile Simulation", *SPIE*, vol. 1927, pp. 413, 1993

[18] M. Fujinaga, N. Kotani, T. Kunikiyo, H. Oda, M. Shirihata, Y. Akasaka, "Three-dimensional Topography Simulation Model: Etching and Lithography", *IEEE Trans. Electron Devices*, vol. 37, pp. 2183, 1990

[19] A. R. Neureuther et al., "Surface-Etching Simulation and Applications in IC Processing", *Proc. Kodak Microelectronics Seminar*, INTERFACE'76, Monterey, CA, pp. 81, 1976

[20] A. Brochet, G. M. Dubroecq, M. Lacombat, "Modelisation des processus d'exposition et de development d'une resine photosensible positive: Application au masquage par projection", *Revue Technique Thompson-CSF*, vol.9, No.2, pp. 285, 1977

[21] W. Henke, M. Weiss, "Three Dimensional Simulation of Reticle Defects in Optical Lithography", *Proc. KTI Microlithography Seminar*, INTERFACE'91, San Jose, CA, pp. 257, 1991

[22] I. Sutherland, R. Sproull, R. Schumacker, "A Characterisation of Ten Hidden Surface Algorithms", *Computing Surveys*, vol. 6, no. 1, pp. 1, 1974

[23] W. Bohmayr, A. Burenkov, J. Lorenz, H. Ryssel, S. Selberherr, "Statistical Accuracy and CPU-time Characteristics of Three Trajectory Split Methods for Monte-Carlo Simulation of Ion Implantation", in: Simulation of Semiconductor Devices and Processes Vol. 6 (eds. H. Ryssel, P. Pichler), Springer Verlag Wien, pp. 492, 1995

[24] J. Lorenz, W. Krüger, A. Barthel, "Simulation of the Lateral Spread of Implanted Ions: Theory", in Proc. NASECODE VI (ed. J.J.W. Miller), Boole Press, Dublin, pp. 513, 1989

[25] H. Ryssel, J. Lorenz, K. Hoffmann, "Models for the Implantation into Multilayer Targets", *Appl. Phys. A*, vol. 41, pp. 201-207, 1986

[26] R.J. Wierzbicki, J.P. Biersack, A. Barthel, J. Lorenz, H. Ryssel, "Reflection Approach for the Analytical Description of Light Ion Implantation into Bilayer Structures", Rad. Eff. and Defects in Solids, 129, 1994

[27] R.J. Wierzbicki, "Analytische Beschreibung der Implantation von Ionen in Ein- und Mehrschichtstrukturen", Ph.D. Thesis, Universität Erlangen-Nürnberg, Verlag Skaker 1994

[28] J. Lorenz, R.J. Wierzbicki, H. Ryssel, "Analytical Modeling of Lateral Implantation Profiles", *Nucl. Instrum. Meth. B*, vol. 96, pp. 168-172, 1995

[29] A. Burenkov, W. Bohmayr, J. Lorenz, H. Ryssel, S. Selberherr, "Analytical Model for Phosphorus Large Angle Tilted Implantation", in: <u>Simulation of Semiconductor Devices and Processes Vol. 6</u> (eds. H. Ryssel, P. Pichler), Springer Verlag Wien, pp. 488, 1995

[30] A. Burenkov, S. List, J. Lorenz, H. Ryssel, "On the Ion Implantation Models for Simulation of FOND Devices", accepted for oral presentation at ESSDERC '95, The Hague, The Netherlands, September 25-27, 1995

[31] K.M. Klein, C. Park, S. Morris, S.-H. Yang, A.F. Tasch, "A Two-dimensional B Implantation Model for Semiconductor Process Simulation Environments", *Nucl. Instrum. Meth. B*, vol. 79, pp. 651, 1993

[32] J. Lorenz, R.J. Wierzbicki, "Efficient Multidimensional Simulation of Ion Implantation into Multilayer Structures", *Proc. 1993 VPAD*, pp. 84, 1993

[33] M. Jacob, P. Pichler, H. Ryssel, R. Falster, "Platinum Diffusion at Low Temperatures", in: <u>Simulation of Semiconductor Devices and Processes Vol. 6</u> (eds. H. Ryssel, P. Pichler), Springer Verlag Wien, pp. 472, 1995

[34] M. Jacob, P. Pichler, H. Ryssel, D. Gambaro, R. Falster, "Determination of Vacancy Concentration in Float Zone and Czochralski Silicon", accepted for oral presentation at ESSDERC '95, The Hague, The Netherlands, September 25-27, 1995

3D TCAD at TU Vienna

E. Leitner, W. Bohmayr, P. Fleischmann, E. Strasser, and S. Selberherr

Institute for Microelectronics, TU Vienna
Gusshausstrasse 27–29, A-1040 Vienna, Austria

Abstract

This paper gives an overview about our research on three-dimensional process simulation. Today's activities are worldwide still suffering from a lack of appropriate geometric modeling, robust gridding, accurate and verifiable physical models as well as computationally efficient numerical algorithms. Possible solutions to some of these problems are demonstrated on the basis of our three-dimensional process simulation tools.

1. Introduction

The development of today's semiconductor devices often requires to investigate three-dimensional problems, where in many cases numerical simulation delivers useful information. For the reliability of such hints the accuracy of the simulation results is crucial. The effects occurring at intrinsic three-dimensional topologies, like corners etc., are gaining importance with shrinking device dimensions. Thus, consideration of the third dimension within the simulation is a must for both process and device simulation.

In contrast to three-dimensional device simulators which are available from universities as well as from commercial sources, the situation in process simulation is quite different: Today it is not possible to perform a complete three-dimensional simulation of a whole process. Impressive work on topography simulation resulted in excellent programs for surface evolution during etching and deposition processes [1] [2]. An engineering workstation is sufficient for this kind of simulations. Also programs for ion implantation based on Monte Carlo methods are available [3] [4] [5]. Where the first versions consumed CPU-times beyond one week (on an HP9000-735), recent developments allow to compute realistic three-dimensional results over night in the amorphous mode, respectively one day for the crystalline mode. Despite these encouraging results, simulation of a whole process fails on missing diffusion and oxidation simulators. Although some three-dimensional simulators have already been presented [6] [7], the complex structures of realistic devices cannot be handled by them because of the lacking grid flexibility.

Therefore development of fully flexible three-dimensional diffusion and oxidation simulators is highly recommended. However, the problems involved are quite complex. Modern devices have more or less arbitrary geometries which are difficult to handle, and the simulation of thermal activated processes requires the solution of coupled, nonlinear partial differential equation systems. This just can be achieved efficiently by use of adaptive gridding techniques as well as highly efficient algebraic methods. Additional challenges lie in the simulation of thermal oxidation, where the diffusion equations and the mechanical equation have to be taken into account. The resulting changes of the geometry reveal high demands to the gridding unit, and additionally the stiff mechanical equations expense the solution of the linear systems. This extremely high complexity within one application has scared of many researchers from tackling the problem.

Besides the numerical and geometrical problems, the quantization of the parameters required by the differential equations accounts for a great deal of controversy. Even for one-dimensional diffusion processes, where the profiles can be measured with satisfying accuracy, the range of the proposed diffusivities is quite large. The situation is even worse for the properties of the point defects: Since it is impossible to measure point defect distributions directly, these are usually quantified due to their effects on the dopant distribution. E.g., the initial distribution is chosen, in order to reach the desired influence on dopant diffusion using a certain set of coupling coefficients. These coupling coefficients have to be determined by calibration of the coefficients, which in turn is influenced by the initial condition. Thus, the missing orthogonality allows to produce almost any result.

Furthermore, it has been pointed out [8], that the lack of multi-dimensional measurement techniques inhibits the calibration of corresponding multi-dimensional simulation tools. On the other hand process simulation seems to be the only possibility to obtain multi-dimensional profiles, because of the insufficient measurement techniques. One may speculate that once we will have reliable one-dimensional physical models, their usage within multi-dimensional simulators will allow to predict doping profiles much more accurately than any measurement technique.

Finally, achieving a complete simulation of a three-dimensional process-flow demands highest flexibility in data management. Data exchanging between different simulators using different data representation forms for geometries (e.g. cellular based, octree based or polygonal based) and profiles stored at different grid types account for a large additional effort which is necessary to couple different simulators effectively. As an example, the generation of a tetrahedral grid for a device geometry computed by a topography simulator such as [1] needs first to convert the cellular based geometry to a polygonal based one. Then the polygonal surface has to be adapted in order to fulfill some conformity conditions and grid points within the solids have to be computed before the tetrahedrization algorithm can be applied. Furthermore, optimization of the grid in terms of element quality and minimum node count requires some kind of optimization loops. All those steps suffer on the enormous amount of data and the structural complexity to be dealt with.

Our work regarding the three-dimensional process simulation resulted in several process simulation modules: In Section 2 we present our simulator for surface evolution and the coupling with physical models for etching and deposition processes. Section 3 deals with the ion implantation module and recent improvements there. In Section 4 we present our module for diffusion processes, and finally, Section 5 contains an outlook on our attempts to tackle three-dimensional oxidation simulation.

2. Topography Simulation

Over time a variety of surface evolution algorithms has been studied to build three-dimensional topography simulators. Among them many algorithms have been reported for resist development in lithography simulation [9] [10] [11] [12], only a few methods have been proposed for the simulation of etching and deposition processes [2] [13] [14] [15]. Basically there are two types of algorithms used for three-dimensional topography simulation. Volume-removal methods divide the material being etched into a large array of rectangular prismatic cells. Each cell is characterized as etched, unetched or partially etched. During etching cells are removed one-by-one according to the local etch rate and the number of cell faces exposed to the etching medium. These algorithms have been successfully used in three-dimensional lithography simulation [9] [10]. Volume-removal methods can easily handle arbitrary geometries, but unfortunately they suffer from inherent inaccuracy, because they favor certain etch directions as was found by many researchers [9] [16]. The second type of surface evolution algorithms represents the surface of the material being etched by using a mesh of points which are connected by line segments to form triangular facets [2] [16]. Depending on the implementation either the mesh points or the facets are moved according to the local etch rates. A mesh management is necessary to maintain the mesh as it moves in time. In general, these algorithms deliver highly accurate results, though with potential topological instabilities such as erroneous surface loops which result from a growing or etching surface intersecting with itself. The surface loops must be located and removed to conserve memory and maintain efficiency of the simulation tool [17].

2.1. A General Method for Surface Advancement

Extensive work in the past has resulted in a general method for surface evolution in three-dimensional topography simulation [1] [18] [19]. This method is based on morphological operations which are performed on a cellular material representation considering the simulation geometry as black and white image (material and vacuum). The resulting surface advancement algorithm allows arbitrary changes of the actual geometry according to a precalculated etch or deposition rate distribution and can support very complex structures with tunnels or regions of material which are completely disconnected from other regions. Surface loops resulting from a growing or etching surface intersecting with itself are inherently avoided.

The material is represented using an array of square or cubic cells, where each cell is characterized as etched or unetched. Additionally, a material identifier is defined for each cell, therefore material boundaries need not be explicitly described as shown in Fig. 1.

The surface boundary consists of unetched cells that are in contact with fully etched cells. Cells on the surface are exposed to the etching medium or to the deposition source, and etching or deposition proceeds on this surface. A linked surface cell list stores dynamically array addresses and rate information of exposed material cells. To advance the surface a structuring element whose spatial dimensions are related to the local etch or deposition rate is applied for the exposed cells. Usually, for anisotropic two-dimensional simulation the structuring element is an ellipse with constant ratio of major to minor axis which is applied in the direction of the local etch or deposition rate vector as shown in Fig. 2, for isotropic movement of the surface point the applied structuring element changes into a circle.

Depending on the simulated process either material cells are removed or added which are located within the structuring element. In case of deposition the structuring

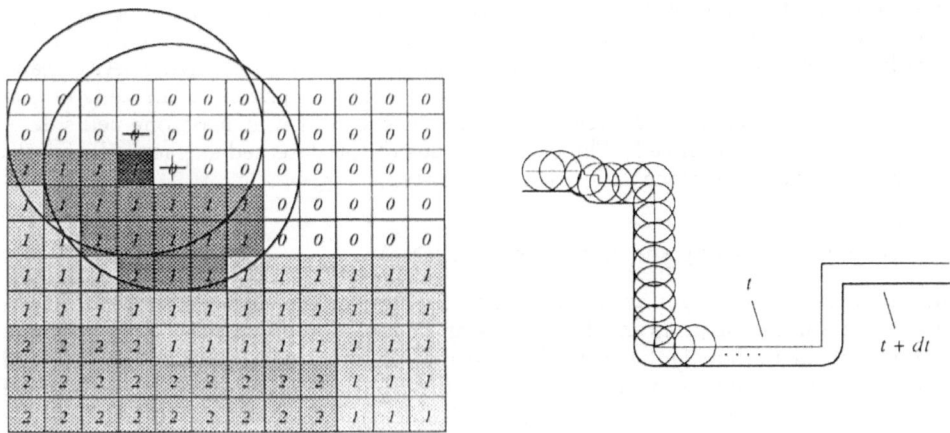

Figure 1: The material representation. The considered surface cell is dark shaded, the number in the cell denotes the material identifier.

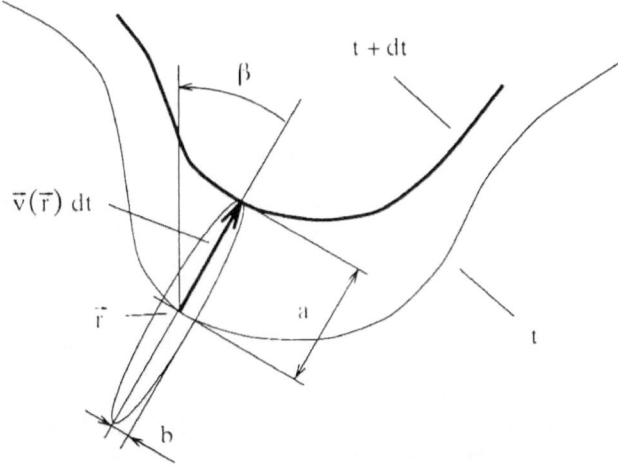

Figure 2: The structuring element for anisotropic surface advancement.

element is centered at the midpoint of the considered surface cell, whereas in case of etching the structuring element is applied at the midpoints of neighboring cells which are located adjacent to the exposed cell sides of the surface cell. For anisotropic three-dimensional surface advancement structuring elements are ellipsoids, for isotropic movement of surface points structuring elements are spheres, although there is no algorithmic restriction on the shape of the applied structuring elements. After each time step the exposed boundary has to be determined. Therefore all cells of the material array are scanned. Material cells are surface cells if at least one cell side is in contact with a vacuum cell. The exposed sides of the detected surface cells finally describe the material surface at a certain time step.

2.2. Modeling of Etching and Deposition Processes

Many topography processes are affected by the shape of the surface. Successful two-dimensional simulation programs for etching and deposition processes use macroscopic point advancement models that consider information about particle fluxes and surface reactions to calculate etch or deposition rate distributions along the exposed surface [20] [21] [22]. This approach is extremely desirable, since a variety of process models for etching and deposition in the literature already exists and quantities such as etch or deposition rates are easily measurable in semiconductor technology.

To determine the rate contributions of incoming particles both in etching and deposition the simulator must be capable to calculate the resulting particle flux incident at a surface point. Therefore a spherical coordinate system with polar angle ϑ and azimuth angle φ is assumed and the region above the wafer is divided up into several surface patches $(N_\varphi \times N_\vartheta)$ as shown in Fig. 3.

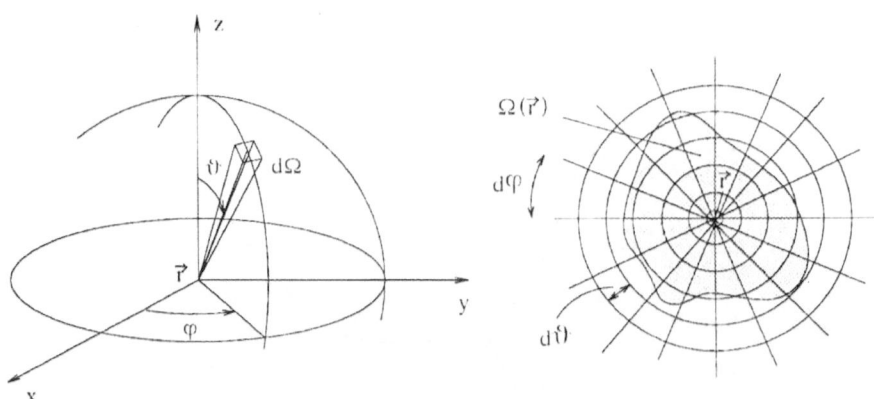

Figure 3: The calculation of the incident particle flux.

The incident flux is then integrated over those patches of the hemisphere which are visible from the surface point $\vec{r}$. To determine if a surface patch is visible from a point on the surface a shadow test has to be performed along a given direction which is within the cellular structure simply the matter of following a discretized line of cells from the surface cell to the boundary of the simulation area. If any cell on this line is a material cell, then the surface cell is shadowed. The calculation of the visible solid angle $\Omega = \Omega(\vartheta, \varphi)$ with $d\Omega = \sin\varphi \, d\varphi \, d\vartheta$ (the radius of the hemisphere may be normalized to one) is then reduced to a series of shadow tests. The number of shadow tests required at a surface point corresponds to the number of patches of the hemisphere. As this number is a constant (typically 90×45 ($N_\varphi \times N_\vartheta$) patches are used), the time required to calculate the visible solid angle for the entire surface is proportional to the number of surface points.

Some processes such as ion milling or crystal etching show a strong dependence on the local surface orientation. The cellular material representation does not provide this information inherently, but the calculation is rather simple. At each exposed side of a surface cell a normal vector can be defined as shown in Fig. 4.

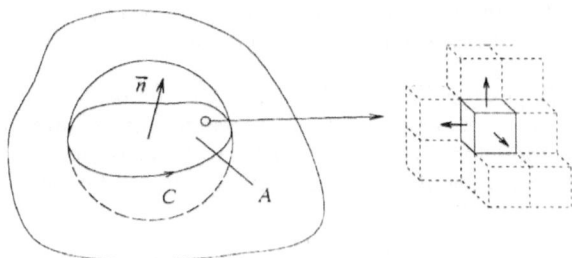

Figure 4: The calculation of the local surface orientation.

The surface normal at a given surface point is then calculated by summing up the normal vectors of surface cells within a certain vicinity to that surface point. For practical simulations surface cells that are located within a sphere are considered, a sphere radius of typically 10 to 15 cells gives highly accurate results.

Etch Models: As a basic concept we consider a linear combination of isotropic and anisotropic reactions of directly and indirectly incident particles to calculate the resulting velocity vector of a surface point. The isotropic reaction is mainly a chemical reaction affected by a reactive gas, in which the reactive particles have short mean free paths and move randomly. The anisotropic reaction is a physical or chemical reaction, where the particles have long mean free paths compared to the device dimensions, and angular particle fluxes must be taken into account. A general process model which accounts for various physical mechanisms like directional etching due to incident ions, etching due to reactive neutrals, and etching caused by reflected or re-emmited particles can be expressed by:

$$v_{iso}(\vec{r}) \;=\; R_{iso}\,, \tag{1}$$

$$v_{dir}(\vec{r}) \;=\; R_i \int_\Omega F_i(\Omega)\, S_i(\alpha) \cos\alpha\, d\Omega + R_n \left[1 + D_i \int_\Omega F_i(\Omega)\, d\Omega \right] \int_\Omega F_n(\Omega) \cos\alpha\, d\Omega +$$

$$+ R_r \int_{H-\Omega} \cos\alpha\, d\Omega\,, \tag{2}$$

with:

$$F_i(\Omega) \;=\; \exp(-\vartheta^2 / 2\,\sigma^2) / N_i\,, \tag{3}$$

$$S_i(\alpha) \;=\; a_1 \cos\alpha + a_2 \cos^2\alpha + a_3 \cos^4\alpha\,, \tag{4}$$

$$F_n(\Omega) \;=\; \cos^m(\vartheta) / N_n\,, \tag{5}$$

$$F_r(\Omega) \;=\; 1\,, \tag{6}$$

where v_{iso} and v_{dir} describe the surface velocity at a surface point along the surface normal, R_{iso} denotes an isotropic etch rate caused by reactive particles of a plasma whose mean free paths are short compared to characteristic device dimensions. The particles are moving randomly, therefore the etch rate has no orientation or flux dependencies. R_i is the etch rate, F_i is the flux distribution, and S_i is the sputter yield due to directly incident ions. R_n and F_n are the etch rate and angular flux distribution for reactive neutrals, D_i is a damage parameter which accounts for the enhancement of the chemical etch rate of neutral particles by the presence of directly incident ions that damage the surface. R_r and F_r describe etching due to reflected particles. α denotes the angle between the incident direction and the surface normal and Ω describes the visible solid angle of the considered surface point.

Deposition Models: Deposition modeling is based on the original work of Blech who developed a model for describing two-dimensional profiles of evaporated thin films over steps [23]. This model is directly applicable to three-dimensional simulation. In three dimensions the components of the growth vector can be calculated by:

$$v_x(\vec{r}) = R_d \int_\Omega F_d(\Omega) \cos \varphi \sin \vartheta \, d\Omega . \tag{7}$$

$$v_y(\vec{r}) = R_d \int_\Omega F_d(\Omega) \sin \varphi \sin \vartheta \, d\Omega , \tag{8}$$

$$v_z(\vec{r}) = R_d \int_\Omega F_d(\Omega) \cos \vartheta \, d\Omega , \tag{9}$$

where R_d denotes the deposition rate on a flat wafer without shadowing and $F_d(\Omega)$ is the angular flux distribution function of incoming particles. A general cosine-based flux distribution function may be expressed [2] as:

$$F_d(\Omega) = \cos^n(A\,\vartheta) / N , \quad for \ \vartheta \leq \pi/2A \ otherwise \ 0. \tag{10}$$

The parameter A restricts the angle of incoming particles, the parameter N allows over-cosine and under-cosine distributions.

2.3. Conversion of the Cellular Geometry Representation

The surface advancement algorithm based on the cellular material representation allows a very stable simulation of arbitrary three-dimensional device structures. Unfortunately, many other simulators can not handle this geometry representation form directly. They require a polygonal geometry representation as input, where the simulation geometry is described by a number of polygons (in most of the cases triangles).

To convert the cellular geometry representation we use the so called Marching Cube Algorithm which was proposed by Lorensen and Cline [24]. This method determines the surface in a logical cube which is created from eight adjacent cells of the material array. According the eight vertices of such a logical cube, there are exact 256 ways a surface can intersect the cube which can be further reduced due to different symmetries to 15 different patterns. Some of the possible patterns are shown in Fig. 5.

The algorithm first determines the surface for one logical cube then moves (or marches) to the next cube. Marching through the whole material array will construct the polygonal geometry representation.

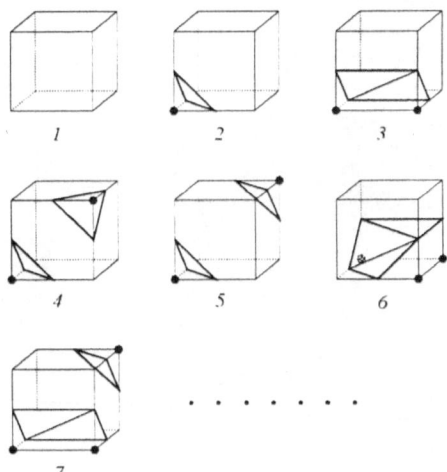

Figure 5: Some of the possible patterns of the Marching Cube Algorithm.

One disadvantage of the Marching Cube Algorithm is that it produces a very large number of triangles which must be reduced afterwards to conserve memory and computational efficiency. We apply a decimation algorithm proposed by Schroeder et al.[25]. In this algorithm multiple passes over all vertices in the mesh are made. During a pass, each vertex is a candidate for removal and, if it meets the specified decimation criteria, the vertex and all the triangles that use the vertex are removed. One such decimation criterion for a vertex is the distance to an average plane which can be calculated using the triangle normals of adjacent triangles to the vertex of interest. The resulting hole after removing the vertex and the triangles in the mesh is patched by a local triangulation. The vertex removal process repeats until some termination condition is met. Usually the termination criterion is specified as a percent reduction of the original mesh.

2.4. An Example

Fig. 6 shows the typical barreling phenomenon which results in ion enhanced plasma etching due to high energetic ions that increase the etch rate where they hit the surface (mainly at the bottom of the trench) and due to reactive neutrals which also attack the sidewalls. The picture also shows the well known aperture effect (etch rate decreasing due to limited delivery of ions and radicals) resulting in a deeper trench where the mask size opening is larger.

The chemical etch rate for this example was $R_n = 0.65\,nm/s$ with $D_i = 6.0$ and the etch time was $400\,s$. The parameter of the particle distribution functions were $\sigma = 2.0$ and $m = 1.0$.

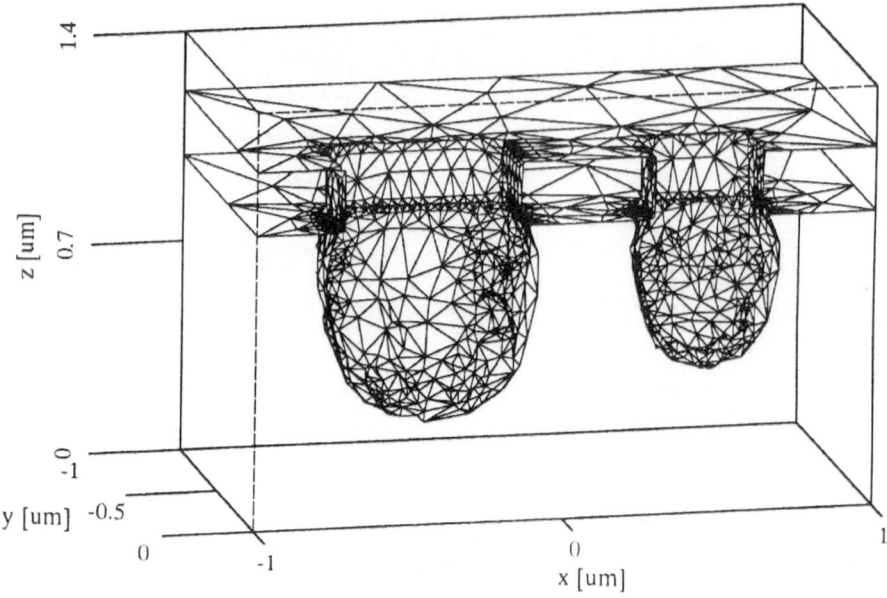

Figure 6: Ion enhanced plasma etching of trenches showing the barreling phenomenon and the aperture effect.

3. Monte Carlo Simulation of Ion Implantation

Introducing controllable amounts of dopant impurities into substitutional sites of a semiconductor crystal predictably modifies its electrical properties. During the past fifteen years, ion implantation has progressed steadily from its initial use as a alternative to diffusion to its present dominating role in the manufacture of VLSI and ULSI circuits. The reason is that we can achieve better control and reproducibility of concentration and depth. Further important features are its flexibility and ability to form almost arbitrary doping profiles, e.g. buried layers, and the favorable factor that ion implantation is a low-temperature process. Since modern annealing methods such as rapid thermal annealing (*RTA*) do not alter the implanted profile very much, the initial profile mainly controls the final result and thus, its determination has become an important task.

The Monte Carlo simulation of ion implantation [4] [5] [26] [27] [28] [29] [30] is rapidly gaining acceptance due to its capability of simulating channeling and damage accumulation phenomena in arbitrary multi-dimensional structures. A well-known disadvantage of the Monte Carlo approach is its considerable demand for computer resources to obtain results with satisfying statistical accuracy.

3.1. Point-Location and Material Detection: The Octree

The Monte Carlo method is based on tracing a large number of trajectories of individual ions on their way through the target until they find their final position. Therefore one crucial aspect of this approach is to determine the spatial location of the ion within the three-dimensional simulation area (*point-location* and *material detection*). To keep the computational effort within reasonable limits we use an octree for discretization of the geometry. This scheme provides a fast solution of the point-location problem by mapping the structure into a hierarchical tree representation [31] [32].

The octree method originates from graphical image processing [33] [34] [35], although in this connection the contrary task, namely to combine areas with the same properties, is desired. Nevertheless this method can be suitably adopted for ion implantation where one big area must be subdivided into smaller zones. In any case the zones shall be as large as possible to achieve the fastest solution.

To meet this requirement the whole geometry is included in one cube (*root cube*). This cube is then subdivided into eight subcubes, if it is not composed entirely of the same material. This procedure is recursively continued for every subcube until either the desired accuracy of the discretization is reached — which is measured by the length of the edges of the cube — or no more intersections of this cube with the polygons defining the target geometry exist (*leaf cube*). At the end each leaf cube is related to exactly one material (Fig. 7).

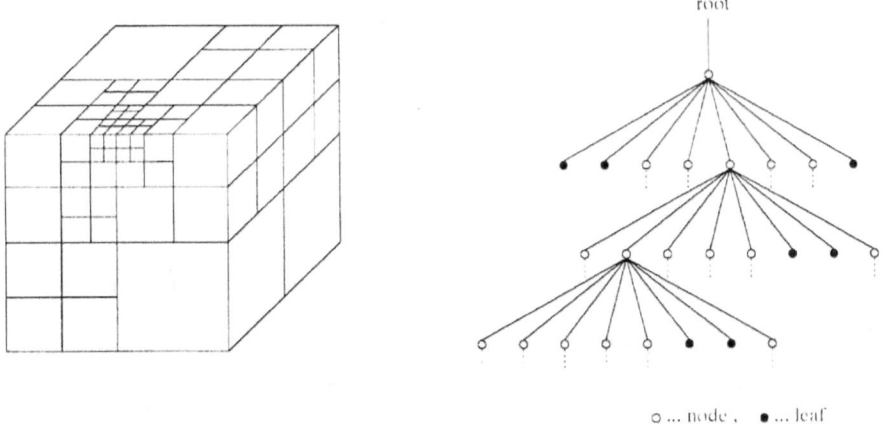

Figure 7: Discretization of a geometry using an octree.

To determine the location of an ion, just a simple test of the coordinate against the related coordinates of the sidewalls of the cube is required, because the cubes of the octree are all aligned with the coordinate axes.

3.2. Amorphous Computation Mode: The Superposition Method

For the Monte-Carlo simulation we let the ions start at equidistant lateral positions (no correlations between trajectories). The number of particles to be simulated is a major concern as the simulation time will be proportional to this quantity. A major part of CPU-time is used for the evaluation of the ion-target interaction. The fundamental idea to reduce the simulation time is to use each ion trajectory several times to determine the history of ions entering the target (Fig. 8). This superposition law holds if the history of all ions is independent as it is the case for amorphous targets [30].

The following algorithm is justified by the superposition law:

1. subdivide the width of the simulation area into N_W subwindows (we use the lateral standard deviation to determine the width of such a subwindow)

2. calculate N *physical* model trajectories in an infinite target for each ion-material combination

3. make copies of this trajectory and move them to corresponding points of each subwindow

4. follow each trajectory copy and check if any boundary is crossed. In this case the used model trajectory is changed according to the new material.

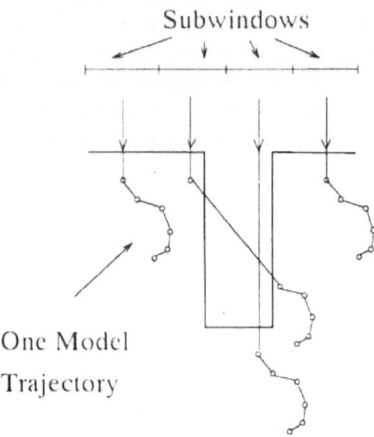

Figure 8: Construction of trajectories from one "physical" ion trajectory.

The simulation will be approximately equivalent to a conventional simulation with $N \cdot N_W$ particles.

However, the computation time required for the necessary geometry checks for the point-location remains uneffected (see Section 3.1).

3.3. Crystalline Computation Mode: The Trajectory Split Method

In order to maintain the performance of ULSI circuits, it is important to form very shallow Source-Drain junctions and to reduce the thermal budgets. Thus, improved models for ion implantation are needed [36] [37] and the traditional assumption of random targets (see Section 3.2) is not longer applicable [38].

The traditional Monte Carlo approach for crystalline targets is based on the calculation of a large number of "distinct" ion trajectories, i.e. each trajectory is usually followed from the ion starting point at the surface of the target up to the stopping point of the ion. Since the majority of ion trajectories ends at the most probable penetration depth inside the structure, the statistical representation of this target region is good. Regions with a dopant concentration several orders of magnitudes smaller than the maximum (in the following we call these areas "peripheral") are normally represented by a much smaller number of ions (typically 10^4 times lower than at the maximum). This results in an insufficient number of events at low concentration areas and leads to statistical noise that cannot be tolerated.

For that reason and inspired by S.-H. Yang [39] we developed the *trajectory split method* [3] [40] for the Monte Carlo simulation of ion implantation on the basis of [41]. Our algorithm drastically reduces the computational effort and is applicable for two and three-dimensional simulations.

The fundamental ideas of our new simulation approach are to locally increase the number of calculated ion trajectories in areas with large statistical uncertainty and to utilize the information we can derive from the flight-path of the ion up to a certain depth inside the target. For each ion, the local dopant concentration C_{loc} is checked at certain points of the flight-path (*checkpoints*). At each checkpoint we relate C_{loc} to the current maximum global concentration $C_{max,current}$ by calculating the ratio $C_{loc}/C_{max,current}$. The result is compared with given relative concentration levels (we define ten levels at 0.3, 0.09, 0.027, ..., 0.3^{10}). Only if the current local concentration falls in an interval below the previous one, a *trajectory split point* is defined at this checkpoint. Therefore our approach is a self-adaptive algorithm because more split points are defined at areas with unsatisfying statistical accuracy. Additional trajectory branches are suppressed, if an ion moves from lower to higher local concentration levels. We store the position of the ion, its energy as well as the vector of velocity and use this data for virtual branches of ion trajectories starting at this split point. In this way, the peripheral areas of the dopant concentration are represented by a much higher number of ion trajectories and the statistical noise is reduced.

Several implementations of this method are conceivable and efficient. We developed three different strategies [40] one shown in Fig. 9. Such a virtual trajectory branch is calculated with the same models and parameters as a regular trajectory, but it starts at the split point with initial conditions obtained from the regular ion. To obtain the correct concentration, a weight is assigned to each branch. The different realizations of the virtual trajectories result from the thermal vibrations of the target atoms [42].

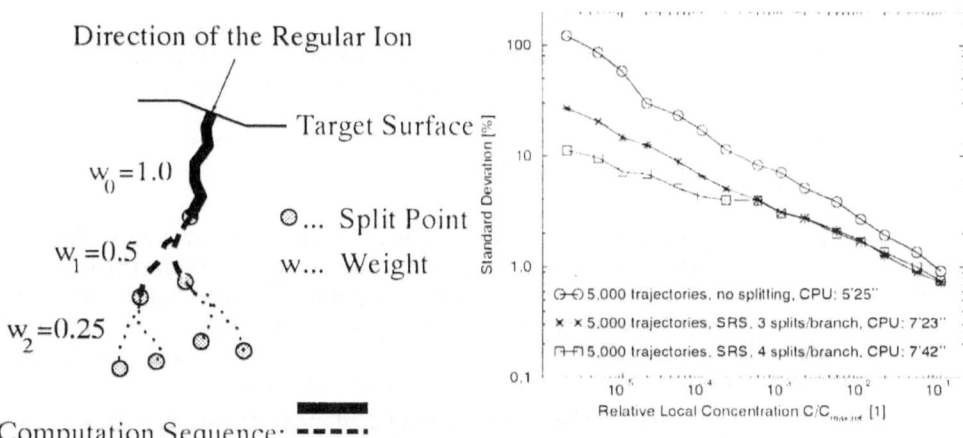

Figure 9: Topological structure of the split-level related split method, the weight of each branch, and the sequence of its calculation

Figure 10: Two-dimensional point response of phosphorus implant, statistical accuracy and CPU time of split-level related split method

To assess the statistical accuracy of the results obtained from the conventional and from the trajectory split methods, we define a mean-square deviation from a reference distribution. For that reason we carry out a conventional simulation with such a high number of ion trajectories (1,000,000) that statistical fluctuations are negligible in the concentration area considered. As an example, we perform a Monte Carlo simulation of a phosphorus implant at 50keV into (100) oriented single-crystal silicon covered by 2.5nm of oxide to obtain point response distributions.

We present the deviation data for the recursive split-level related split method [40] in Fig. 10 calculated with 5,000 distinct ion trajectories. The relative concentration in this figure is defined as the ratio of $C/C_{max,ref}$, where $C_{max,ref}$ means the maximum concentration of the reference distribution. The computational effort is approximately proportional to the number of distinct ion trajectories and the additional overhead due to trajectory splits is only 25% to 35%.

Further important advantages of the trajectory split method are its lower sensitivity to the local concentration and the opportunity to individualize its error behavior. Increasing the number of splits per each branch, cf. Fig. 10, and/or initializing more than one virtual branch at each split point leads to a significantly smaller error in peripheral areas without effecting the statistic in other regions. In other words there is a chance of optimizing the relation between CPU time and required statistical accuracy for a particular problem.

It should be mentioned that our new strategy is also best suited to compute the collision cascade of a displaced target atom ("recoil"). Depending on the ion energy and the atomic mass ratio of the ion and the recoil some collisions cause a considerable number of recoils which lead to a statistical "over-representation" of such events. The new method offers the possibility to optimize the recoil statistic by a random deletion of recoil trajectories at such places and by splitting them at peripheral areas of the collision cascade.

3.4. An Example

The Source-Drain doping in minimum-size transistor designs is an intrinsically three-dimensional problem. Furthermore, in modern shallow-junction processing channeling may affect the device performance [43]. Thus, an ion implantation into a field oxide corner of a conventional LOCOS structure is best suited to demonstrate the merits and the applicability of the *superposition method* and the *trajectory split method*, respectively.

For the simulations we used a phosphorus implant of $5 \cdot 10^{13} \text{cm}^{-2}$ at 40keV. Fig. 11 shows the geometry of the conventional LOCOS. The screening oxide thickness is 10nm and the ion beam was tilted for $-7°$ in the xy-plane. To investigate the channeling effects we cut the LOCOS geometry by a horizontal xz-plane 10nm below the silicon/silicon-dioxide interface. Fig. 12 and Fig. 13 show the amorphous and crystalline mode simulation results for the conventional LOCOS.

From these results follows that in the active region the dopant concentration near the silicon surface is significantly decreased whereas the doping at the periphery (along the bird's beak) remains uneffected due to the dechanneling property of the thicker oxide in that region ($\approx$ 60nm).

The required computational effort for such a rigorous three-dimensional simulation is approximately proportional to the exposed area (*implantation window*) and depends on the energy of the ions. On a HP 9000-735/100 workstation our example takes about 15 hours using the *superposition method* and about one day using the *trajectory split method*. Compared to conventional strategies the speed-up is about a few orders of magnitudes for the amorphous mode and about five for the crystalline mode.

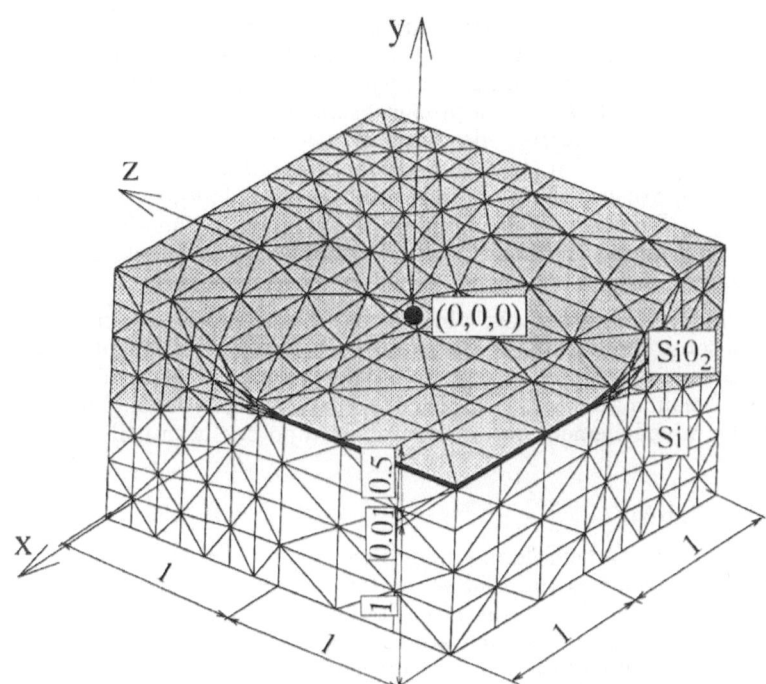

Figure 11: Corner of the conventional LOCOS structure

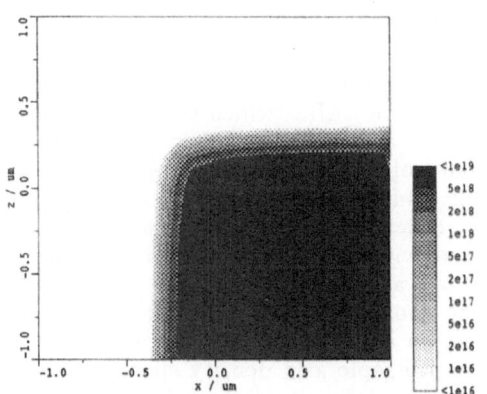

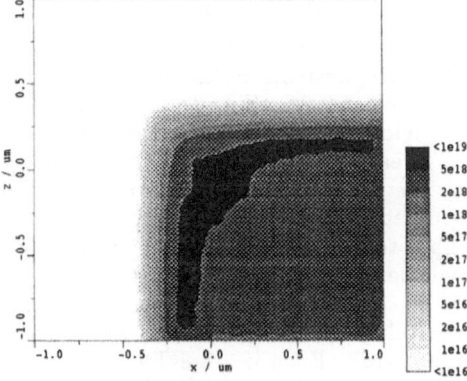

Figure 12: Concentration of phosphorus in cm^{-3} 10nm below the silicon surface (conventional LOCOS geometry is cut by a horizontal xz-plane. amorphous mode)

Figure 13: Concentration of phosphorus in cm^{-3} 10nm below the silicon surface (conventional LOCOS geometry is cut by a horizontal xz-plane. crystalline mode)

4. Diffusion Processes

The solution of the coupled nonlinear differential equations coming up from diffusion processes usually is done by means of the Finite Element Method respectively the Finite Box Method. Both require a grid, which has to resolve the geometry and the doping profiles in order to achieve a proper discretization. This grid plays the keyrole for solution efficiency, because it determines both the condition and the size of the sparse equation systems. For three-dimensional simulations the size of the systems is the major hurdle, because the numerical solution effort follows in the very best case approximately $n^{3/2}$ with n as the number of unknowns. Therefore the grid has to be optimized in order to fulfill the required accuracy requirements using a minimum number of nodes.

We divide our gridding activities into two parts: generation of an initial grid and adaptation of the grid according to the changed accuracy requirements as the diffusion advances. For initial grid generation we have developed a tetrahedrization module, based on the Delaunay criterion. In order to adapt the grid throughout the diffusion simulation the mixed-element decomposition method has been utilized [44].

4.1. Grid Generation

Grid generation plays a keyrole in three-dimensional process and device simulation. The amount of data forces the grid generator to keep the optimal balance between accuracy and efficiency. The geometry has to be partitioned into the smallest number of elements, which still allows accurate solving of the governing equations. These elements have to fulfill certain quality requirements and the gridding process should work entirely automatically. Another challenge lies within a typical TCAD situation, where many simulators have to interact. Tools might generate their own grids which are not valid as an input for subsequent tools. In order to avoid unnecessary re-interpolation of attributes defined on a grid and to minimize gridding efforts after each simulation step the ideal grid generator should also have the capability to read and modify existing grids. For instance, a grid output from one tool has to be made geometry conform before it can be input to the next tool [45].

We use the concept of a "place nodes and link" algorithm [46], which in our opinion offers the best means to deal with the above mentioned tasks. Grid nodes first have to be placed according to local and global grid densities, after which they are linked to yield the grid elements. Note that this is the only approach which allows the input of already existing grid nodes. Thus, grid nodes of different grids can be merged, or previously generated grids can be refined by adding additional grid nodes. The geometry can be arbitrarily complex, the boundary points of the geometry are a direct input to the grid generator. Conformity of the grid with the boundary of the geometry will be discussed in the next section. Many disadvantages of octree-based methods which intrinsically place the grid nodes can be avoided. For instance, the limited flexibility in node placement and the sensitivity to alignment of the geometry with the octree. Even if 2-4-8-Trees [47] are used, anisotropic grid density specifications cannot generally be fulfilled. Finally, the need for special gridding techniques near boundaries to avoid staircase-like representations of the geometry makes the use of octree-based methods less favourable.

After the linking step, the resulting tetrahedral grid elements have to meet the desired quality requirements. There are two degrees of freedom how one can change this element quality. Both *node placement* and the *type of tetrahedrization* have a crucial influence. Unfortunately, they are not entirely independent of each other, thus, an optimization loop becomes necessary. After the tetrahedrization, the node placement might have to be changed or grid nodes might have to be inserted and the tetrahedrization process has to be repeated. Even if the tetrahedrization process is in some sense optimal (e.g. Delaunay tetrahedrization) elements of poor quality cannot be avoided beforehand. In three dimensions the *aspect ratio* of a simplex can be defined as the ratio of the radii of the circumscribed sphere to the inscribed sphere [48]. Typical elements with poor aspect ratios are [49]:

Needle: A tetrahedron with a very long and a very short edge.

Cap: A tetrahedron, where the radius of the circumsphere is much larger than the longest edge.

Sliver: A tetrahedron consisting of four nearly coplanar points, which are evenly spaced on a great circle of the circumsphere.

Especially in two dimensions interesting dependencies between various optimization criteria (no large angles, no small angles, height) and the Delaunay triangulation have been shown [48]. Essentially, *Steiner points* are added to the initial point set (*Steiner triangulation*) and the Delaunay triangulation of the modified point set is used. Considering a fixed point set in two dimensions the Delaunay triangulation is known to maximize the minimum angle. In three dimensions the Delaunay tetrahedrization (DT) minimizes the maximum radius of a minimum-containment sphere [50]. The minimum-containment sphere is the smallest sphere that contains the tetrahedron. It can be identical to the circumsphere. The need for a general, numerically robust tool to compute the DT for a fixed point set becomes evident. In fact, the DT is the only tetrahedrization which is dual to the well known Voronoi diagram. Thus, if the Box integration method is applied, the DT becomes a *necessary* tool to avoid negative control volumes.

4.1.1. Boundary Conformity

If a general tetrahedral Delaunay grid is required to be conform with the boundary, the boundary has to be represented by a surface triangulation where each triangle fulfills the Delaunay criterion. If and only if at least one empty sphere passing through the three points of a boundary triangle exists, the boundary triangle is a Delaunay triangle. This sphere can have any size. For the Box integration method a stronger criterion has to be satisfied.

Boundary refinement criterion: The smallest sphere passing through the three points of a boundary triangle may not contain any other point.

The step in which the input boundary triangles (generally not Delaunay) are modified according to this criterion is called *boundary refinement*. This is a complex two and a half dimensional problem (Literature mostly covers the one and half dimensional case, [45][48][51][52]). Multiple domains pose no difficulties. The boundary refinement step has to be applied to the interfaces as well. Our boundary refinement module uses a combination of flipping boundary triangles (sometimes also called edge-swapping) and inserting additional boundary points to satisfy the above stated criterion: If two adjacent boundary triangles are coplanar and violate the criterion, their common edge can be flipped without having to insert a point. This technique greatly reduces the number of additional inserted points. (A discussion of such local transformations can be found in [53].) Inserting points only becomes necessary along edges of the geometry or in the case of (parallel) planes with small distance compared to the size of the triangles.

4.1.2. Delaunay Tetrahedrization

We implemented an incremental tetrahedra construction algorithm (Fig. 14). It can be imagined as "advancing front" that pervades the input geometry. The domain on the backside of the front is entirely tetrahedrized and the domain on the front side not at all. The great advantage of this algorithm is how rigorous it deals with complex input geometries (e.g. multiple connected). It only tetrahedrizes the interior of the geometry, because the advancing front will be stopped by the surface triangulation of the boundary. (As opposed to other algorithms, where a convex hull is tetrahedrized and a subsequent segmentation step is necessary which distinguishes interior and exterior elements.)

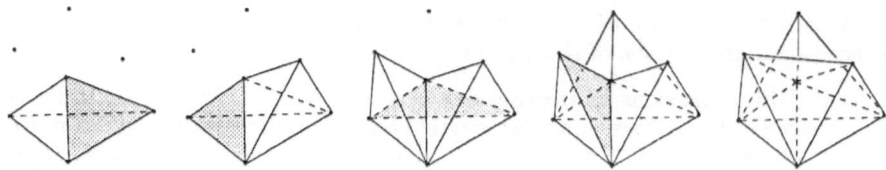

Figure 14: Incremental algorithm

A typical problem for DT algorithms are degeneracies due to cospherical point sets. (If more than four points are located on the perimeter of an empty sphere, the point set is said to be cospherical.) The implemented algorithm allows any number of cospherical points and will not have to add points to deal correctly with these cases.

Another difficulty lies in finite-precision arithmetics. Topological point connectivity evolves from numerical calculations. Thus, numerical errors manifest in topological inconsistencies. The degree of freedom in choosing the point connectivity is spent to satisfy the Delaunay criterion. In order to guarantee a topological correct tessellation the Delaunay criterion need not always be satisfied. A small tolerance ϵ has to be granted. Sugihara and Iri discussed this topic for the two-dimensional Voronoi diagram in [54]. Our implementation solves this problem. It detects topological inconsistencies that would occur due to a sole adherence to the Delaunay criterion and overrides the criterion by ϵ. The algorithm uses an octree point location method and runs in $O(n \log n)$ time (Fig. 15).

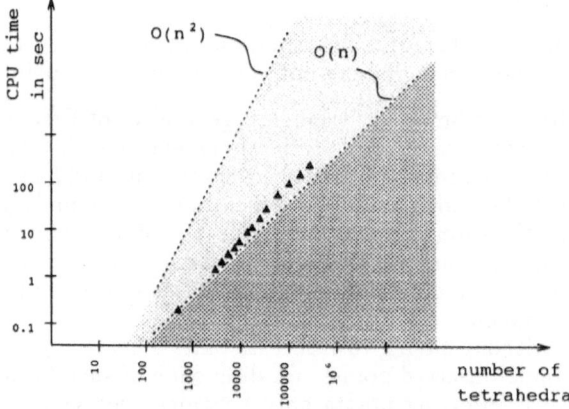

Figure 15: Time complexity (HP 9000-735/100)

So far the point connectivity describing the linking of the grid nodes was of concern. However, in the three-dimensional space the *face connectivity* plays an important role. One face (triangle) connects the two spaces on each side. In the presence of cocircular points in three dimensions (more than three points which are located on an empty circle) the two half spaces on each side of the plane containing the cocircular points can be connected via triangles in more than one way. If on one side of the plane the set of cocircular points is triangulated in a different manner than on the other side, the result is inconsistent face connectivity. Note that the cocircular point set implies the existence of two neighboring sets of cospherical points. The question arises whether it is possible to tetrahedrize a cospherical point set, if its convex hull contains a fixed and given triangulation of cocircular points. Fig. 16 depicts a case where the specified face connectivity cannot be achieved by any tetrahedrization of the interior. This is a similar problem to the un-tetrahedrizable polyhedron mentioned in [48]. It can be solved by local transformations of the tetrahedra, which are connected to the plane containing the cocircular point set.

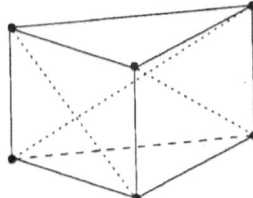

Figure 16: Impossible face-connectivity

4.2. Adaptive Gridding Strategy

Once an initial grid is available which resolves the geometry, it has to be adapted to the implanted doping profiles and the adaptation has to be redone as the diffusion advances and the local discretization errors exceed their limits. Several grid updates are necessary throughout a whole simulation. Thus, a fast adaptation algorithm is needed in order to keep the computational overhead for the grid management low. Therefore, we use a recursive element decomposition method.

For recursive refinement algorithms it is indispensable to preserve the grid quality, i.e., the unavoidable degradation of the grid quality has to stay within a limit which is independent of the number of refinements. If we can find a refinement method which keeps the element shape, the above requirement is fulfilled automatically. Unfortunately for tetrahedra no such method exists. However, it is possible to define a two-level splitting method, which preserves the element shape during multiple refinement.

We divide a tetrahedron into four tetrahedra of the same shape and one octahedron. The four tetrahedra are located at the parent's corners and the remaining part has octahedral shape (Fig. 17). An octahedron we divide into six octahedra of the same shape and eight tetrahedra. The six octahedra are located at the parent's corners and the remaining parts have tetrahedral shape (Fig. 18). In order to discretize an octahedron, we split it into eight tetrahedron, each of which has one face of the octahedron as ground plane and the octahedral center as opposite node.

The first refinement step introduces elements with a new aspect ratio. The elements generated by all following refinement steps have either the shape of the tetrahedra or

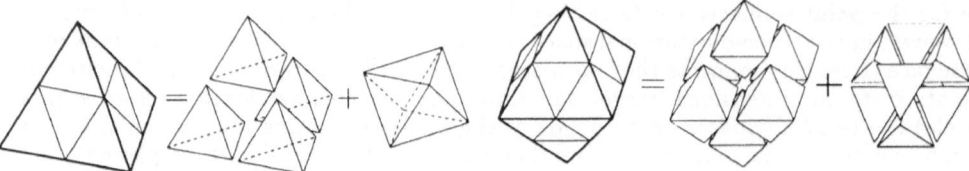

Figure 17: Tessellation for a tetrahedron   Figure 18: Tessellation for an octahedron

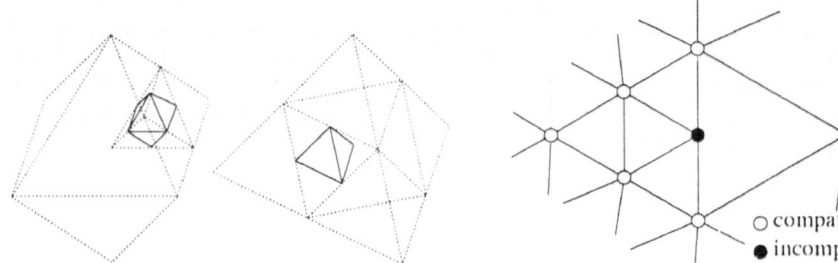

○ compatible node
● incompatible node

Figure 19: Shape preservation for recur-   Figure 20: Incompatible elements
sive refinement

the shape of the octahedra which exist after the first refinement step (see Fig. 19). Thus, the element quality is affected only at the first refinement step. Taking into account the discretization of the octahedron permits a reasonable comparison of the octahedral child of the tetrahedron: we compare the element quality of the tetrahedral parent with the element quality of the tetrahedra used for the discretization of the octahedron. It can be shown, that the degradation of the element quality according (11) is limited to a factor of $1/2$. A similar comparison of the tetrahedral child with the octahedral parent results in a maximum degradation factor of $1/4$.

$$Q = \frac{V}{h_{max}^3} \tag{11}$$

As the refinement is always done locally, unrefined elements may be adjacent to refined ones. These neighboring elements are called incompatible elements, and we define the order of incompatibility as the difference of the refinement levels of two adjacent elements. In our algorithm the order of incompatibility is restricted to one. A two dimensional example of such an incompatible situation is shown in Fig. 20. In order to estimate the grid quality at a compatible node between incompatible elements, we use (12), where V_i are the volumes of all elements incident to the node [55]. It can be shown, that the degradation of this nodal grid quality is limited to a factor of $1/4$ for the tetrahedron and $1/8$ for the octahedron.

$$Q = \frac{\min(V_i)}{\max(V_i)} \tag{12}$$

4.3. Discretization Method

For the practical application of the mixed-element decomposition method, we implemented a finite element method in order to solve (13), where C_k are the quantities to solve and J_k are the according fluxes. The quantities are coupled by the coefficients

γ_{kl}, which allow modeling of generation and recombination, and by the coefficients α_{kl}, which couple the fluxes of the different quantities.

$$\frac{\partial C_k}{\partial t} + \mathrm{div}\, J_k + \sum_{l=1}^{N_Q} \gamma_{kl} = 0$$

$$J_k = \sum_{l=1}^{N_Q} (a_{kl} \cdot \mathrm{grad}\, C_l)$$

$$\text{for} \quad k = 1, \ldots, N_Q \tag{13}$$

As discretizing element we use the tetrahedron with linear shape functions. The octahedrons coming from the mixed-element decomposition method are split into eight tetrahedrons for discretization. The weak form of (13) results from the method of Galerkin-weighted residuals [56] and is integrated by means of Gaussian integration using one integration point in the center of the tetrahedron.

In order to estimate the discretization errors, we use a gradient smoothing method [56]. This method uses the shape functions of the elements for a continuos approximation of the piecewise constant gradient of the solution. By twofold integration of the gradient difference along an element the local dosis error of the solution is computed. The decision about local grid refinement respectively coarsening is based on a weighted combination of the local dosis error related to the local dosis and the local dosis error related to the global dosis. The weights allow to control the grid density at high concentration levels nearly independently from the grid density at low concentration levels.

The discretization in time uses the standard finite differences method according the Backward-Euler scheme. Error estimation is used in order to test the accuracy of the previous time step as well as to predict the size of the next time step. The estimation is based on a parabolic approximation of the piecewise linear evolution of the solution at each node. By means of extrapolation the size for the new time step is estimated.

4.4. Solution Strategy

For solution of the nonlinear equation systems, we implemented a damped Newton iteration scheme for the coupled equations. To achieve quadratic convergence, we extend the element matrices to the full Fréchet derivative of the nonlinear functional. The resulting linear equation systems are solved iteratively by means of a BICGStab-solver [57] with an incomplete Gauß-elimination for preconditioning. For a particular time step the initial condition for the Newton iteration scheme is obtained by quadratic extrpolation of the solution of the previous time steps. This technique reduces the number of Newton iterations by one for each time step.

Once the grid is adapted to the initial doping profiles, it is successively modified after each time step according to the adaptation criteria. All elements are checked upon their discretization error and are either refined or replaced by their parents. On inserting of new nodes, the solution values are interpolated by a third order interpolation function, which satisfies the continuity condition for the gradients.

4.5. An Example

To demonstrate the benefits of the adaptive gridding algorithm, we computed a diffusion step for a Boron channel-implant in a conventional LOCOS-structure at 1000 °C and an annealing time of 30min. Figure 11 shows the coarse initial tessellation of the simulation region where the field oxide is on top of the silicon bulk. The channel implant has been computed by Monte-Carlo ion implantation [3] with an energy of 20keV and a dose of 1e14cm^{-2}. The initial grid has been adapted to the initial profile for a discretization error limit of 1% relative to the total implanted dosis. The grid for the silicon region with the distribution is shown in Fig. 21 and consists of 9534 nodes and 19259 elements.

As the diffusion advances the steep gradients are smoothed and therefore, the grid was reduced continuously by the automatic adaptation algorithm. Thus, the final grid at the end of the simulation consists of only 8093 nodes and 15636 elements (Fig. 22).

For the needed 34 time steps the program consumed a CPU-time of 41 minutes on an HP 9000-735/100 workstation and used approximately 32MB of memory, which shows, that fully three-dimensional diffusion simulation with a controlled discretization error is feasible.

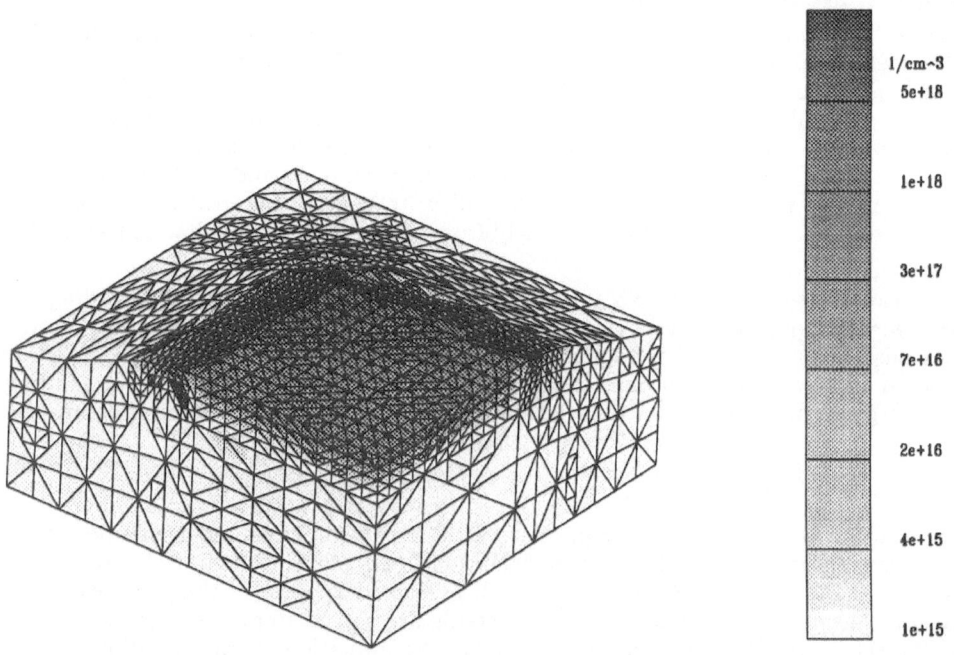

Figure 21: The implanted Boron profile in the Silicon region

5. Oxidation

The physics of thermal oxidation results in a high number of coupled differential equations (typically 6-10 for a three-dimensional simulation). Even for two-dimensional simulations the equation systems resulting from a coupled solution are very large. The standard methods using Newton methods or decoupled iteration schemes combined with iterative solvers for the linear systems require significant amounts on computational resources. Therefore, we investigated other (preferably matrix-free) algorithms and found the multigrid method feasible for oxidation problems.

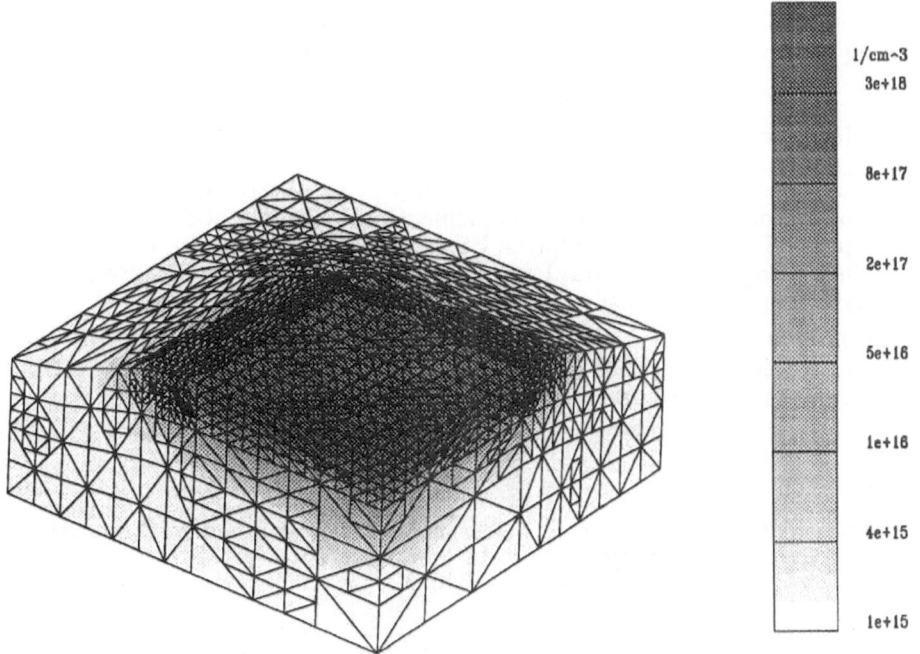

Figure 22: Boron profile after 30min. annealing at 1000 ° C

We are currently evaluating the efficiency of multigrid methods for mechanical equations, and the results seem very promising. We compared different methods in order to solve the mechanical stress/strain equations, including the nonlinearity coming up from large displacement. For a problem with about 16000 unknowns, where the iterative solver diverged on solving the linear system, the Gaußsolver consumed about 2 min CPU time and 60 MB memory on an DEC-Alpha workstation for one matrix inversion within the Newton scheme which needed 5 iterations, resulting in 10 min overall CPU time. The multigrid algorithm was able to solve the whole nonlinear problem within about 30 sec and needed just 8 MB memory.

Another advantage of the multigrid method is, that the solution effort is growing just linearly with the number of nodes. This feature makes the multigrid method appearing best suited for use in three-dimensional process simulation with coupled stress/strain equations, where high node counts are unavoidable.

6. Conclusion

Despite the large variety of problems coming up from three-dimensional process simulation, several steps can already be simulated reasonably. Topography simulation of etching and deposition processes and ion implantation simulation deliver accurate results and consider the physical effects sufficiently. The development of diffusion and oxidation process simulation tools is still at the beginning. However, we have shown reasonable solutions to the grid adaptation and grid generation problem, which allows us to perform simulation of dopant diffusion with a controlled discretization error and reasonable demands on computational resources.

Furthermore, the coupling of different simulators is a must. The problems arising from different data representation formats have been pointed out and solutions for converting polygonal to cellular based geometries and vice versa have been shown.

Acknowledgement

TU Vienna wants to acknowledge important support by Austria Mikro Systeme AG, Unterpremstätten, Austria; Digital Equipment Corp., Hudson, USA; Hitachi Ltd., Tokyo, Japan; LSI Logic Corp., Milpitas, USA; Motorola Inc., Austin, USA; National Semiconductor Corp., Santa Clara, USA; Philips B.V., Eindhoven, The Netherlands; Siemens AG, München, Germany; Sony Corp., Atsugi, Japan.

Part of this work was carried out in cooperation between PROMPT (JESSI project BT8B) and ADEQUAT (JESSI project BT11) and has been funded by the EU as ESPRIT projects No. 8150 and 8002, respectively.

References

[1] E. Strasser, K. Wimmer, and S. Selberherr. A New Method for Simulation of Etching and Deposition Processes. In *1993 International Workshop on VLSI Process and Device Modeling*, pp 54–55, 1993.

[2] E.W. Scheckler and A.R. Neureuther. Models and Algorithms for Three-Dimensional Topography Simulation with SAMPLE-3D. *IEEE Transactions on Computer-Aided Design of Integrated Circuits*, 13:219–230, 1994.

[3] W. Bohmayr, G. Schrom, and S. Selberherr. Trajectory Split Method for Monte Carlo Simulation of Ion Implantation Demonstrated by Three-Dimensional Poly-Buffered LOCOS Field Oxide Corners. In *Int.Symposium on VLSI Technology, Systems, and Applications*, Taipei, 1995.

[4] A.M. Mazzone and G. Rocca. Three-Dimensional Monte Carlo Simulations – Part I: Implanted Profiles for Dopants in Submicron Devices. *IEEE Trans.Computer-Aided Design*, CAD-3(1):64–71, 1984.

[5] A.M. Mazzone. Three-Dimensional Monte Carlo Simulations – Part II: Recoil Phenomena. *IEEE Trans.Computer-Aided Design*, CAD-4(1):110–117, 1985.

[6] S. Odanaka, H. Umimoto, M. Wakabayashi, and H. Esaki. SMART-P: Rigorous Three-Dimensional Process Simulator on a Supercomputer. *IEEE Trans.Computer-Aided Design*, 7(6):675–683, 1988.

[7] C. S. Sun, O. K. Kwon, C. G. Hwang, and H. J. Hwang. Three-Dimensional Numerical Simulation for Low Dopant Diffusion in Silicon. In S. Selberherr, H. Stippel, and E. Strasser, editors, *Simulation of Semiconductor Devices and Processes*, pp 413–416. Springer-Verlag Wien New York, 1993.

[8] M. E. Law. Challenges to Achieving Accurate Three-Dimensional Process Simulation. In S. Selberherr, H. Stippel, and E. Strasser. editors, *Simulation of Semiconductor Devices and Processes*. pp 1–8. Springer-Verlag Wien New York, 1993.

[9] E.W. Scheckler, N.N. Tam, A.K. Pfau, and A.R. Neureuther. An Efficient Volume-Removal Algorithm for Practical Three-Dimensional Lithography Simulation with Experimental Verification. *IEEE Transactions on Computer-Aided Design of Integrated Circuits*, 12:1345–1356, 1993.

[10] W. Henke, D. Mewes, M. Weiß, G. Czech, and R. Schließl-Hoyler. A Study of Reticle Defects Imaged into Three-Dimensional Developed Profiles of Positive Photoresists Using the SOLID Lithography Simulator. *Microelectronic Engineering*, 14:283–297, 1991.

[11] Y. Hirai, S. Tomida, K. Ikeda, M. Sasago, M. Endo, S. Hayama, and N. Nomura. Three-Dimensional Resist Process Simulator PEACE (Photo and Electron Beam Lithography Analyzing Computer Engineering System). *IEEE Transactions on Computer-Aided Design of Integrated Circuits*, CAD-10:802–807, 1991.

[12] T. Ishizuka. Bulk Image Effects of Photoresist in Three-Dimensional Profile Simulation. *Journal for Computation and Mathematics in Electrical and Electronic Engineering*, 10:389–399, 1991.

[13] K.K.H. Toh, A.R. Neureuther, and E.W. Scheckler. Algorithms for Simulation of Three-Dimensional Etching. *IEEE Transactions on Computer-Aided Design of Integrated Circuits*, 13:616–624, 1994.

[14] J. Pelka. Three-dimensional Simulation of Ion-Enhanced Dry-Etch Processes. *Microelectronic Engineering*, 14:269–281, 1991.

[15] S. Tazawa, F.A. Leon, G.D. Anderson, T. Abe. K. Saito. A. Yoshii, and D.L. Scharfetter. 3-D Topography Simulation of Via Holes Using Generalized Solid Modeling. In *Int.Electron Devices Meeting*, pp 173–176. 1992.

[16] K.K.H. Toh. *Algorithms for Three-Dimensional Simulation of Photoresist Development*. PhD thesis, University of California, Berkeley. 1990.

[17] J.J. Helmsen, E.W. Scheckler, A.R. Neureuther, and C.H. Séquin. An Efficient Loop Detection and Removal Algorithm for 3D Surface-Based Lithography Simulation. In *Workshop on Numerical Modeling of Processes and Devices for Integrated Circuits*, pp 3–8, 1992.

[18] E. Strasser, G. Schrom, K. Wimmer, and S. Selberherr. Accurate Simulation of Pattern Transfer Processes Using Minkowski Operations. *IEICE Transactions on Electronics*, E77-C:92–97, 1994.

[19] E. Strasser and S. Selberherr. Algorithms and Models for Cellular Based Topography Simulation. *IEEE Trans.Computer-Aided Design*. accepted for publication, 1995.

[20] W.G. Oldham, A.R. Neureuther, C. Sung, J.L. Reynolds, and S.N. Nandgaonkar. A General Simulator for VLSI Lithography and Etching Processes: Part II - Application to Deposition and Etching. *IEEE Trans.Electron Devices*, ED-27:1455–1459, 1980.

[21] J. Lorenz, J. Pelka, H. Ryssel, A. Sachs. A. Seidel. and M. Svoboda. COMPOSITE - A Complete Modeling Program of Silicon Technology. *IEEE Transactions on Computer-Aided Design of Integrated Circuits*. CAD-4:421–430. 1985.

[22] S. Tazawa, S. Matsuo. and K. Saito. Unified Topography Simulator for Complex Reaction including both Deposition and Etching. In *1989 Symposium on VLSI Technology*, pp 45–46, 1989.

[23] I.A. Blech. Evaporated Film Profiles Over Steps in Substrates. *Thin Solid Films*, 6:113–118, 1970.

[24] W.E. Lorenson and H.E. Cline. Marching Cubes: A High Resolution 3D Surface Construction Algorithm. *Computer Graphics*. 21:163–169, 1987.

[25] W.J. Schroeder, J.A. Zarge, and W.E. Lorenson. Decimation of Triangle Meshes. *Computer Graphics*, 26:65–70, 1992.

[26] M.T. Robinson and O.S. Oen. Computer Studies of the Slowing Down of Energetic Atoms in Crystals. *Physical Rev.*, 132(6):2385–2398, 1963.

[27] M.T. Robinson and I.M. Torrens. Computer Simulation of Atomic-Displacement Cascades in Solids in the Binary-Collision Approximation. *Physical Rev. B*, 9:5008–5024, 1974.

[28] J.P. Biersack and L.G. Haggmark. A Monte Carlo Computer Program for the Transport of Energetic Ions in Amorphous Targets. *Nucl.Instr.Meth.*, 174:257–269, 1980.

[29] J.F. Ziegler, J.P. Biersack, and U. Littmark. The Stopping and Range of Ions in Solids. Pergamon Press, 1985.

[30] G. Hobler and S. Selberherr. Monte Carlo Simulation of Ion Implantation into Two- and Three-Dimensional Structures. *IEEE Trans.Computer-Aided Design*, 8(5):450–459, 1989.

[31] H. Stippel and S. Selberherr. Three Dimensional Monte Carlo Simulation of Ion Implantation with Octree Based Point Location. In *Int. Workshop on VLSI Process and Device Modeling (1993 VPAD)*, pp 122–123, Nara, Japan, 1993. Jap.Soc.Appl.Phys.

[32] H. Stippel and S. Selberherr. Monte Carlo Simulation of Ion Implantation for Three-Dimensional Structures Using an Octree. *IEICE Transactions on Electronics*, E77-C(2):118–123, 1994.

[33] W.D. Fellner. *Computergrafik*. Reihe Informatik. B.I. Wissenschaftsverlag, 1992.

[34] M. Mäntyla. *An Introduction to Solid Modeling*. Computer Science Press, 1988.

[35] S. Abramowski and H. Müller. *Geometrisches Modellieren*. B.I. Wissenschaftsverlag, Mannheim, 1991.

[36] K.M. Klein, C. Park, and A.F. Tasch. Monte Carlo Simulation of Boron Implantation into Single-Crystal Silicon. *IEEE Trans.Electron Devices*, 39(7):1614–1621, 1992.

[37] G. Hobler, H. Pötzl, L. Gong, and H. Ryssel. Two-Dimensional Monte Carlo Simulation of Boron Implantation in Crystalline Silicon. In W. Fichtner and D. Aemmer, editors, *Simulation of Semiconductor Devices and Processes*, volume 4, pp 389–398, Konstanz, 1991.

[38] H. Ryssel, G. Prinke, K. Haberger, K. Hoffmann, K. Müller, and R. Henkelmann. Range Parameters of Boron Implanted into Silicon. *Applied Physics A*, 24:39–43, 1981.

[39] S.-H. Yang, D. Lim, S. Morris, and A.F. Tasch. A More Efficient Approach for Monte Carlo Simulation of Deeply-Channeled Implanted Profiles in Single-Crystal Silicon. In *Int. Workshop on Numerical Modeling of Processes and Devices for Integrated Circuits NUPAD V*, pp 97–100. Honolulu, 1994.

[40] W. Bohmayr, A. Burenkov, J. Lorenz, H. Ryssel, and S. Selberherr. Statistical Accuracy and CPU Time Characteristic of Three Trajectory Split Methods for Monte Carlo Simulation of Ion Implantation. In *Simulation of Semiconductor Devices and Processes*, Erlangen, 1995.

[41] A. Phillips and P.J. Price. Monte Carlo Calculations on Hot Electron Tails. *Appl.Phys.Lett.*, 30(10):528–530, 1977.

[42] M. Jaraiz, J. Arias, E. Rubio, L.A. Marques, L. Pelaz, L. Bailon, and J. Barbollan. Dechanneling by Thermal Vibrations in Silicon Ion Implantation. In *X International Conference on Ion Implantation Technology*, pp Abstract P–2.19, 1994.

[43] W. Bohmayr, G. Schrom, and S. Selberherr. Investigation of Channeling in Field Oxide Corners by Three-Dimensional Monte Carlo Simulation of Ion Implantation. In *Int.Conference on Solid-State and Integrated-Circuit Technology*, Beijing, 1995.

[44] E. Leitner and S. Selberherr. Three-Dimensional Grid Adaptation Using a Mixed-Element Decomposition Method. In *Simulation of Semiconductor Devices and Processes*, Erlangen, 1995.

[45] S. Halama. *The Viennese Integrated System for Technology CAD Applications — Architecture and Critical Software Components.* PhD thesis, Technische Universität Wien, 1994.

[46] V. Srinivasan, L. Nackman, J. Tang, and S. Meshkat. Automatic Mesh Generation Using the Symmetric Axis Transformation of Polygonal Domains. *Proc.IEEE*, 80(9):1485–1501, 1992.

[47] P. Conti. *Grid Generation for Three-Dimensional Semiconductor Device Simulation.* Hartung-Gorre, 1991.

[48] M. Bern and D. Eppstein. Mesh Generation and Optimal Triangulation. In F.K. Hwang and D.-Z. Du, editors, *Computing in Euclidean Geometry*, pp 201–204. World Scientific, 1992.

[49] T.J. Baker. Element Quality in Tetrahedral Meshes. *7th Int. Conf. on Finite Element Models in Flow Problems*, Huntsville, Alabama, 1989.

[50] V.T. Rajan. Optimality of the Delaunay Triangulation in R^d. *Proc. 7th ACM Symp. Comp. Geometry*, pp 357–363, 1991.

[51] L. Nackman and V. Srinivasan. Point Placement for Delaunay Triangulation of Polygonal Domains. *Proc. 3rd Canadian Conf. Comp. Geometry*, pp 37–40, 1991.

[52] A. Saalfeld. Delaunay Edge Refinements. *Proc. 3rd Canadian Conf. Comp. Geometry*, pp 33–36, 1991.

[53] B. Joe. Three-Dimensional Triangulations from Local Transformations. *SIAM J.Sci.Stat.Comput.*, 10(4):718–741, 1989.

[54] K. Sugihara and M. Iri. Construction of the Voronoi Diagram for One Million Generators in Single-Precision Arithmetic. *Proc.IEEE*, 80(9):1471–1484, 1992.

[55] R.E. Bank. *PLTMG: A Software Package for Solving Elliptic Partial Differential Equations*, volume 7 of *Frontiers in Applied Mathematics*. SIAM, 1990.

[56] O.C. Zienkiewicz and R.L. Taylor. *The Finite Element Method.* McGraw-Hill, fourth edition, 1989.

[57] H.A. van der Vorst. BI-CGSTAB: A Fast and Smoothly Converging Variant of BI-CG for the Solution of Nonsymmetric Linear Systems. *SIAM J.Sci.Stat.Comput.*, 13(2):631–644, 1992.

Multi-Dimensional TCAD: The PROMPT/DESSIS Approach

M. Westermann[a], T. Feudel[b], N. Strecker[b],
S. Gappisch[b], A. Höfler[b]and W. Fichtner[a,b]

[a]ISE Integrated Systems Engineering AG,
Gloriastrasse 35, CH-8092 Zürich, SWITZERLAND
[b]Swiss Federal Institute of Technology,
Integrated Systems Laboratory, Zürich, SWITZERLAND

Abstract

Designing new semiconductor devices for very large scale integrated circuits
(VLSI) requires intensive use of process and device simulation tools to reduce development costs. However, valid device simulation results can only be
achieved when a high geometrical modeling precision has been reached during
the process simulation phase. Accurate finite element simulators embedded in
a multi-dimensional process simulation environment help to fulfill this quality
requirement. Since general-purpose three-dimensional (3D) process simulators
are not yet available, a modern design environment combines solid modeling
techniques with one-dimensional (1D) and two-dimensional (2D) finite element
simulators. In this article we present a consistent and integrated process simulation environment that eases the characterization and optimization of semiconductor devices. The straightforward modeling in all three dimensions of an
EEPROM cell illustrates the presented approach.

1. Introduction

Process simulation involves a comprehensive study of the physical phenomena ruling
the behavior of semiconductor material under various external influences. Despite
of the complexity of these physical processes, modern 1D and 2D process simulators
are accurate enough to make their use cost-effective compared to experiments based
on repeated device fabrication and result analysis. The application area of process
simulators ranges from simple process flow visualization to highly difficult tasks like
the generation of geometrical models for evaluating the electrical behavior of the
device.

While 1D and 2D process simulators are widely used, the development of 3D process
simulators is far from meeting the minimal requirements of the designer community.
The reason of this development delay is obviously the complexity of the geometrical
computations resulting from the movement of material boundaries; a problem which
is more easy to solve in 1D and 2D. Furthermore, the difficulty of mapping 1D and
2D physical models into the third dimension should not be underestimated. This
task can get rather complex when one thinks about the chemical, mechanical and

electrical effects involved in device fabrication. A more practical problem is inherent to all 3D process simulators using finite element methods to achieve high simulation precision: the enormous simulation time required by 3D simulators forces the user to look for alternative simulation methods whenever accurate 3D models are not absolutely necessary. Fortunately, solid modeling is an appropriate alternative for building general semiconductor devices. We have chosen this approach in order to complete our simulation environment. Although solid modelers cannot reach the accuracy of finite element simulators, their application domain covers a large area of the process simulation field.

A simulator management tool becomes necessary as soon as various process simulators are used in the same environment. No user would like to convert a process flow manually into the different languages required by the process simulators. However, a management system is more than a simple file conversion tool. It provides the user with a database containing the appropriate simulation models for each process step. Furthermore, the communication between the simulators—mainly the exchange of geometrical and doping data—is also handled on this level.

After a presentation of the process simulation management system developed in our laboratory, the systematic use of the different simulators according to their application domain will be outlined. Section 3 briefly explains the EEPROM process technology used for our example. While many semiconductor devices can be simulated in the way shown here, the EEPROM device reflects typical problems encountered at the design phase of a VLSI device. Sections 4 to 6 illustrate the modeling procedure applied to the EEPROM process. The resulting simulated geometrical structures and doping distributions allow an early evaluation of the device's quality. This evaluation step is then completed by performing a set of device simulations.

2. Process Simulation Management

2.1. The Simulation Environment

In order to understand the integration of multi-dimensional process simulators within the semiconductor design environment, this section gives an overview of the simulation tools used in our laboratory. The presented program structure is general enough to reflect process simulation environments implemented in other laboratories. Fig. 2.1 shows the main components of such a software system. Three main levels can be recognized in the figure:

The **process simulation** level contains accurate process simulators of various dimensionality and functionality. Starting from the basic layout and process information contained in a Semiconductor Process Representation (SPR) file, a system called LIGAMENT [1] generates automatically input files for process simulators of various dimensionality. Currently the process simulators TESIM [2] and DIOS [3] and the solid modeling system PROSIT [4] are connected to LIGAMENT. The generality of the environment, however, allows other simulators to be plugged in rather easily.

The **structure modeling and grid generation** level accomplishes the difficult task of transforming meshes generated by process simulators into device simulation grids. The grid generators MDRAW [5] and OMEGA [6] provide an interface between the process and device simulation levels.

Finally, the designer can observe on the **device simulation** level the electrical behavior of its semiconductor device with the multi-dimensional device and circuit simulator DESSIS [7]

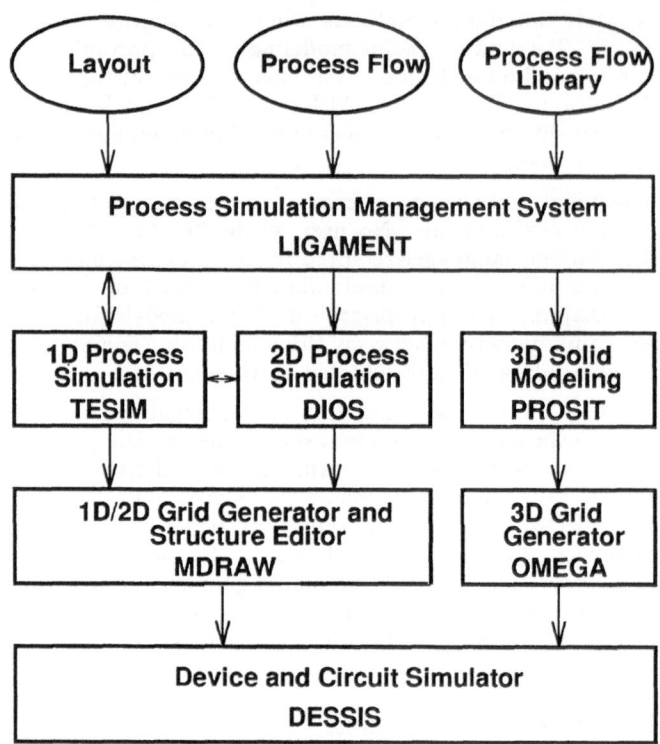

Figure 1: The semiconductor process simulation environment used at the ETH-IIS laboratory

2.2. Managing Process Simulators

In the environment shown in Fig. 2.1 LIGAMENT is the central package of the process simulation level. It reads the layout data and process flow description provided by the user and generates input files for the process simulators TESIM and DIOS, and for the solid modeler PROSIT. An obvious advantage of this modeling environment is the unique layout and process flow files the user must provide. The knowledge of the input syntax of the various simulators included in this environment is not necessary in order to run the simulators.

Another concern of a management tool is the concept of simulation model abstraction. Specifying the right simulation parameters is by no means obvious because of the required simulator specific knowledge. LIGAMENT overcomes this restriction by providing the user with a process flow library which has been tuned to support a predefined set of simulation models. This library has been developed by an expert of the process simulation tools embedded in our environment. Optionally, an automatic generation of these library functions from manufacturing information minimizes the risk of introducing modeling errors due to the data transfer between the technology and the simulation sites. In the ideal case, the process library is directly created from the wafer fab control files.

As shown in Fig. 2.1 LIGAMENT reads also 1D simulation results. The returned TESIM results are used to complete the input file for PROSIT. A direct connection between TESIM and DIOS allows the user to start a simulation with less expensive 1D simulations as long as no mask edges appear in the simulation domain. The first patterning operation which introduces 2D features in the selected simulation domain produces an automatic switching from TESIM to DIOS.

2.3. Semiconductor Process Representation

Since LIGAMENT generates automatically input files for various process simulators, the main goal of such a system is to provide a unique process description interface to the designer. A Semiconductor Process Representation (SPR) language becomes essential for the definition of the process flow. The SPR used in LIGAMENT is based on a block-structured programming language like C. While the syntax is not the most important design issue—an interactive graphical user interface is currently under development—the SPR language must provide the designer with a set of commands that allows a concise definition of the process flow. In any case, a post-editing of the generated simulator input files should never be necessary or even wished. An example of the SPR syntax is shown below:

```
defop buox (time:          37 min,
                temperature:   1000 degC,
                oxygen:        100 %,
                include_ramps: unknown,
                fast_process:  unknown)
{
  if (!$fast_process) {

    if ($include_ramps) {
      anneal(time: 10 min, temp: 900 degC 1000 degC,
             oxygen: 0.2 1/min, nitrogen: 5 1/min)
    }

    anneal(temp: $temperature, time: 15 min,
           oxygen: 0.2 1/min, nitrogen: 5 1/min)
    ...
  }

  if ($fast_process) {
    deposit(material: oxide, thickness: 40 nm)
  }
}


* calling the function:
buox (time: 50 min, include_ramps: NO,
      fast_process: YES)
```

The function above is used to model a buffer oxide. Depending on the flags "include_ramps" and "fast_process", the designer can specify the simulation accuracy of the complete process ("fast_process" flag) and the approximation level of single process steps ("include_ramps" flag). In the example above, the buffer oxide is approximated with a deposition operation.

Different types of commands are implemented in the SPR language. With the command sets shown below, the designer has the full control over the generated input files:

Execution control commands. These operations are used to control the interpretation of the SPR file. The user can define his own functions, and he can use the few branching statements provided by the language ("if" and "else" control structures). This branching mechanism is mainly used to simplify the process flow in order to reduce the final simulation time. Time consuming oxidation process steps can be replaced, for example, with simple oxide deposition commands, provided that no implantation steps are preceding the oxidation.

LIGAMENT control commands. These commands are used to control LIGAMENT itself. The target simulator and a set of options can be specified at this command level.

Layout definition commands. These commands associate the geometry stored in a layout file with a mask suitable for a photolithographic process. A photo-mask can also be defined as a Boolean combination (union, subtraction, intersection, or inversion) of other masks.

Simulator commands. This type of operations are simulator control commands. They do not belong to the process flow itself, but are required by almost all currently existing simulators. One example is the "title" command which initializes the simulator, or the "region" command which defines the simulation domain. A special command called "insert" allows the insertion of simulator specific commands between normal process flow operations. The example below is a typical example of an initialization sequence containing simulator commands. The "insert" command is used to control the graphical output of DIOS. The keyword "ngra=1000" means that a picture will be printed after each process step.

```
title(text: "EEPROM Simulation")

* Define the cutline:
region(x0: 5um, y0: -11um,
       x1: 5um, y1:  17um)

* Set some Dios graphic flags:
insert(dios: "repl(cont(ngra=1000,
               print=1))")
```

Process commands. The classical process steps are available in the SPR language: "anneal", "deposit", "epitaxy", "etch", "implant" and "pattern".

Specific simulator attributes (etching rates or other model parameters) can be defined with the "type" command. This command allows the user to group attributes into labeled records. The identifier in the "type" command is used as a reference for all process commands which need simulator dependent attributes. Any etching operation with the same "type" reference (plasma_poly for example) will have the same attributes. This mechanism allows the user to separate simulator specific attributes from the process flow.

```
* Written in the process flow library:
type(name:    plasma_poly,
      dios:    "rate(mat=po, a0=400,
                     mat=ox, aniso=5),
               over=50")


* Written in the user provided process
* flow description:
etch(material: poly, thickness: 400 nm,
      type: plasma_poly)
```

The example above shows the definition procedure of simulator dependent etching rates. The plasma etching rates are specified in DIOS as a set of simulation parameters: polysilicon is etched at a rate of 400 nm per minute (a0=400), while oxide is removed at a rate of 5 nm/min. An overetching time of 50% allows the complete removal of the exposed polysilicon.

For the rest of this paper, we used an EEPROM cell as a representative example for our simulation environment. We have chosen this example because it reflects a realistic situation from an actual project between our group and a semiconductor wafer fab.

3. The EEPROM Process

Fig. 3 shows the final result of the process simulation in two dimensions. The EEPROM cell is a two-transistor device. The select transistor gives access to the memory cell. Programming and erasing of the cell is performed with Fowler-Nordheim tunneling through the thin tunnel oxide. During programming a high voltage is applied on the control gate while the drain is at ground. The strong electric field over the tunnel oxide causes electrons to tunnel to the isolated floating gate resulting in a threshold shift. The cell can be similarly erased with a high voltage on the drain and control gate at ground.

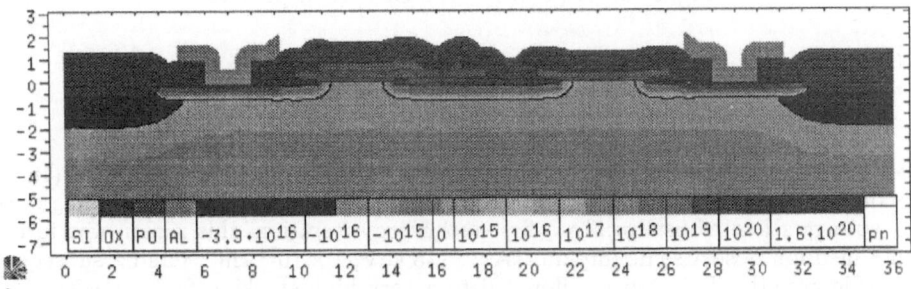

Figure 2: Simulated cross section of the two-transistor EEPROM cell

In the actual implementation, a conventional 2-μm CMOS technology has been modified to implement the EEPROM cell. The process sequence and mask layout are shown in Fig.3 After n-well formation, patterning of active areas, channel stop implant and field oxide growth, the arsenic tunnel implant windows are opened in the nitride layer covering the active areas. The tunnel implantation is followed by a thermal oxidation in a wet ambient (350 nm tunnel implant oxide). After removing the nitride layer, the first gate oxide is grown. Then 10 nm of tunnel oxide are grown in a hole etched into the 350 nm thermal oxide currently present in the tunnel region. Following the deposition of the first polysilicon layer ("floating" and "select" gates), a traditional CMOS process is used to complete the device.

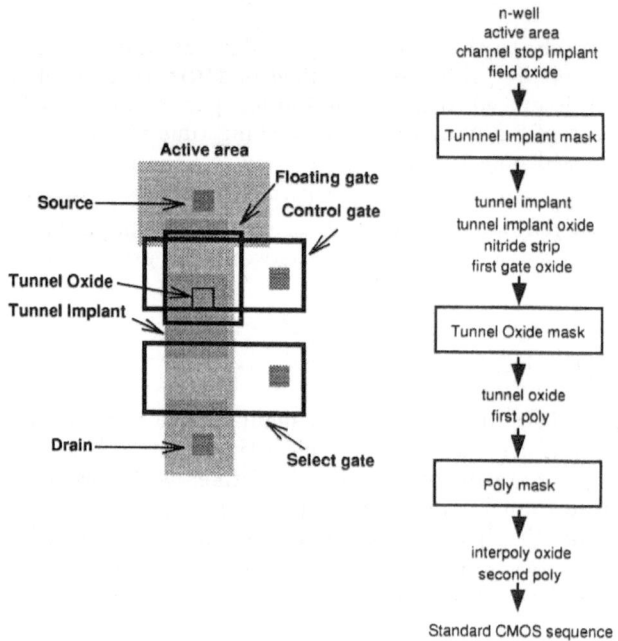

Figure 3: Sequence of the EEPROM process

4. 1D Process Simulation with Tesim

Of critical importance for the device behavior is the n-type doped region (tunnel implant) for the carrier injection into the floating gate and for the drain doping of the high-voltage select transistor. In a first technology iteration, this region was formed with a medium dose arsenic implant (As^+, 175 keV, 1×10^{14} cm^{-2}) and a subsequent wet oxide growth. This process flow led to a very high stacking fault density. First simulations resulted in a bad agreement between measurement and simulation for the sheet resistance of the tunnel implant region which was two times too high. This could be corrected by decreasing the arsenic segregation coefficient from 0.1 to 0.01. But nevertheless, the simulation of such problems is always critical because of a combination of the channeling effect, oxidation enhanced diffusion, doping enhanced oxidation, Fermi-level dependent diffusion and segregation.

Process sequence	Measured Rs	Sim. w/o channeling	Sim. with channeling
1000 °C, N$_2$, 20 min	464	481	438
1000 °C, N$_2$, 20 min 950 °C, H$_2$O$_2$, 100 min	1023	2053	1173
950 °C, H$_2$O$_2$, 100 min	1183	2612	1282

Table 1: Comparison of measured and simulated sheet resistance values Rs (all Rs values in $\Omega/\square$)

For a better understanding of the problem we performed some special experiments (Table 1). The SIMS results showed us that the as-implanted profile was dominated by a strong channeling tail. This profile could be reproduced with our simulator for ion implantation processes CRYSTAL-TRIM [8]; a module currently included in TESIM. Fig. 4 shows a comparison of the measured profile in (100) silicon with simulation results for implantation into amorphous and crystalline silicon. The exact description of the as-implanted profile is especially important for the simulation of processes with low thermal budget or a strong dopant loss during oxidation, as in our case. A comparison of TESIM simulation results with measured data (Fig. 5) shows a very good agreement after using the correct implantation profile that precedes the high temperature treatment. This is also true for the sheet resistance. These simulations have been performed with default parameter for the steady-state diffusion models. The initial fitting of the segregation coefficient was the wrong way.

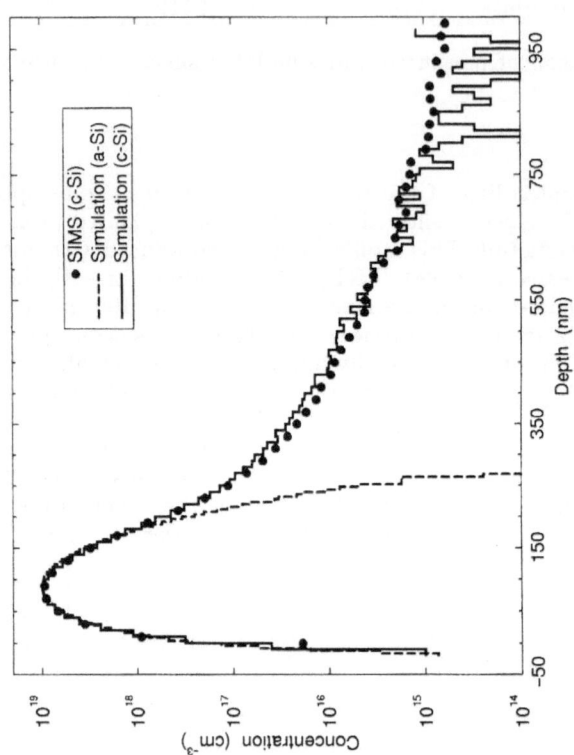

Figure 4: Measured and simulated profiles after an arsenic implantation (175 keV, 1×10^{14} cm^{-2}, 7° tilt, no rotation) through 20 nm SiO$_2$

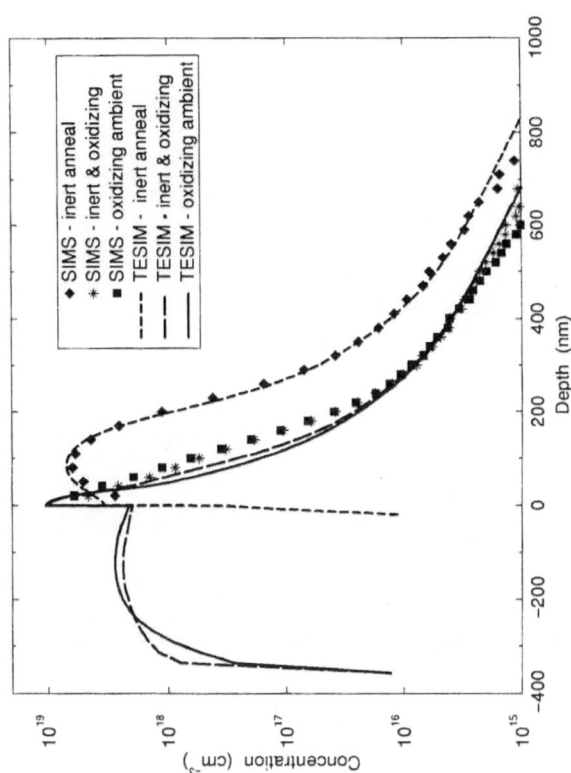

Figure 5: Simulated and SIMS arsenic profiles in the tunnel implant region

The utilization of a hierarchical concept as explained in this situation also shows the power of our approach. Using LIGAMENT and the SPR environment, it is straight-forward to have a hierarchical model for a particular process step; in this case ion implantation.

5. 2D Process Simulation with Dios

The 2D simulations performed with DIOS have been used to observe the influence of the photolithography masks on the device structure. Different simulation domains have been placed on the layout shown in Fig. 3. One simulation topic was the device modifications resulting from a maximum mask shift tolerance of 0.5 μm. From this procedure, we figured out the following design error: The mask used for patterning the tunnel implant areas protected the nitride at the field oxide edges (Fig. 6 top part). Thus, the tunnel implant oxide was prevented from growing and a gap separating this oxide and the field oxide appeared after some overetching steps. The full dose of the drain implant could penetrate due to this gap, and the breakdown voltage of the select transistor was reduced (Fig. 6 bottom part).

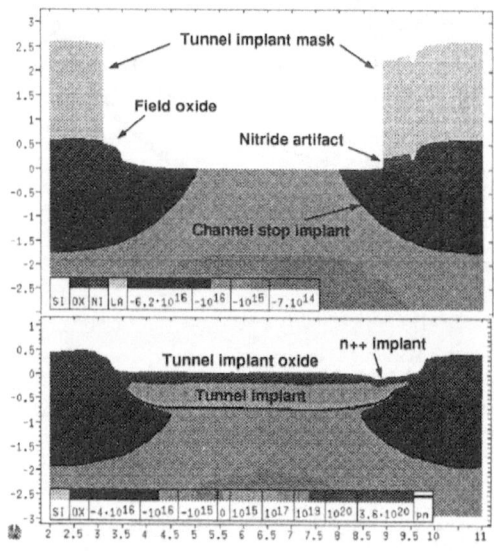

Figure 6: The influence of a mask shift on the device structure. The top picture shows the gap on the right side, and the bottom picture shows the modified drain implantation profile

Another application of DIOS for the EEPROM modeling is the optimization of the tunnel implant dose. The breakdown voltage of the "select" transistor should be as high as possible in order to stand the high programming voltages. On the other hand, the erase coupling factor which is also influenced by the tunnel implant should not be significantly decreased. Fig. 7 shows an example for the optimization of the breakdown voltage of the "select" transistor. In this example, the breakdown occurs due to avalanche generation at the drain corner.

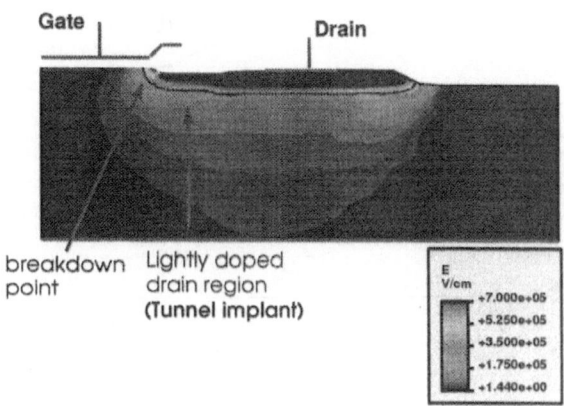

Figure 7: Breakdown simulation of the "select" transistor

Fig. 8 displays the breakdown voltage in dependence of the dose of the tunnel implant. In this simulation a critical electrical field of $0.75*10^6$ V/cm (indicating first occurrence of band-to-band tunneling) was used as a criteria for the breakdown. During the measurements a current of 10 μA was used to detect the breakdown voltage. We estimated that an optimal value of 2×10^{13} cm^{-2} will be necessary for the next technology iteration.

6. 3D Solid Modeling with Prosit

Experience with Constructive Solid Geometry (CSG) packages has shown that the reliability of a solid modeler is crucial in VLSI process simulation. The key problem of solid modeling techniques is the robustness of Boolean operations performed on two solids. An intersection calculation has a high probability to fail when numerical errors resulting from the limitation of floating point arithmetics are ignored in the implementation of geometric algorithms. For this reason, a cell-based approach has been chosen for the realization of PROSIT. The geometrical data structure uses a regular subdivision of the 3D simulation domain into blocks of equal size [4] Each block can be refined into a group of cells by splitting it along the x, y or z axis. Modeling operations are performed by letting an ellipsoidal brush walk along material interfaces and modify the traversed layer structure. This principle of the painting brush is already widely used in computer graphics for drawing the border of polygons and for painting lines. We have extended this concept into the third dimension,

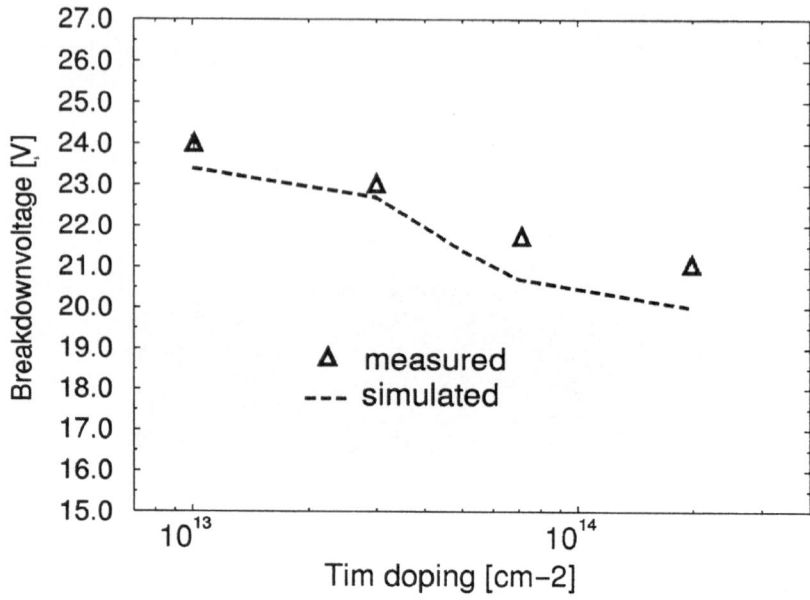

Figure 8: Simulated and measured breakdown voltage of the "select" transistor

and we have increased the modeling flexibility by giving the 3D painting brush any rotational symmetric shape. This modeling technique allows both robust geometrical operations and the inclusion of external 1D and 2D simulation results.

The EEPROM cell has been modeled in 3D by applying a sequence of different modeling brushes on the geometrical structure of the device. As an example, Fig. 9 shows two types of oxidation modeling brushes. Oxidation is often accompanied by stress effects resulting from a thin nitride mask that reduces the area of silicon reacting with oxygen. Since oxidation produces expansion of material, the resulting profile can have a complex shape which might only be foreseen by accurate process simulators. It often becomes necessary to include precomputed or measured profiles. However, a few classical oxidation techniques can also be defined analytically in a solid modeler. The left side of Fig. 9 shows a LOCOS example modeled with a "bird's beak" brush. The brush has been defined analytically with the functions proposed by Guillemot et al. in [10].

The example shown on the right of Fig. 9 requires one modeling brush for each polysilicon gate oxidation. The brush defines the amount of polysilicon transformed into oxide. The center of the brush is slightly above the ellipsoid's center-of-mass. In this way, the amount of transformed material is higher on the upper side of the gate than on the lower side which gets less oxygen from the surrounding atmosphere. Unfortunately, this modeling method does not take into account the shadowing effect produced by the "control" gate located on top of the "floating" gate. This second gate behaves like a oxygen mask for the lower gate. A more precise modeling requires a location dependent interpolation of two brushes and a better definition of the brush

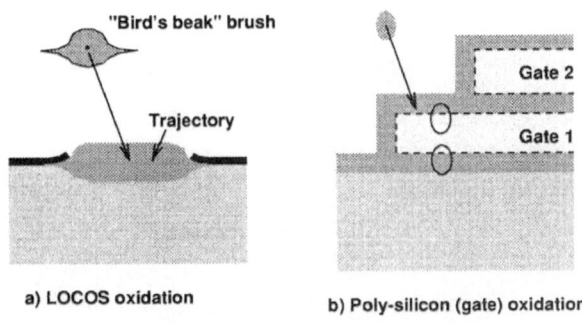

Figure 9: LOCOS and polysilicon oxidation examples

trajectory. However, the EEPROM cell has been modeled with a unique brush for each gate.

The solid modeling operations defining the EEPROM structure have been generated with LIGAMENT. A few 1D and 2D process simulations have been performed in order to optimize the modeling operations given to PROSIT. Instead of performing a set of oxidation operations on the polysilicon gates, we prefer to include the 1D and 2D simulation results in the deposited polysilicon and oxide layers. A 2D DIOS simulation through the gate allowed us to resize the gate mask in order to include the consumed polysilicon. 1D TESIM simulations have been mainly used to correct the gate oxide and polysilicon layer dimensions.

A block size of one micron and a cell size of one third of a micron have been chosen for the simulation domain resolution used during the process flow's interactive observation. This resolution is too low when a correct geometry is expected for further device simulations. Thus, a block size of 200 nm and a subdivision of a block into 4×4×20 cells give an accurate device structure. A simplified version of this structure—the metal layer has been removed and the oxide layer reduced to a minimum—is used as a starting geometry for the device simulation. Fig. 10 shows the electrostatic potential resulting from an "erase" command applied to the double gate regions.

7. Conclusions

The presented simulation environment allowed us to characterize an EEPROM cell in a systematic way: 1D simulations were used for tuning the implantation process steps. Experimental verifications allowed us to validate the simulation results. 2D simulations showed us the importance of mask adjustments and also allowed us to optimize the tunnel implant dose. Finally, PROSIT has been used to build a 3D structure suitable to 3D device simulation. The fast visualization capability of PROSIT also helped us to better understand the 3D structure of this device. However, the central application domain of PROSIT resides in the generation of initial geometries for performing 3D device simulations.

One of the main advantages of our process simulation management system is the agreement of the models used in the process simulators. Therefore we could easily use

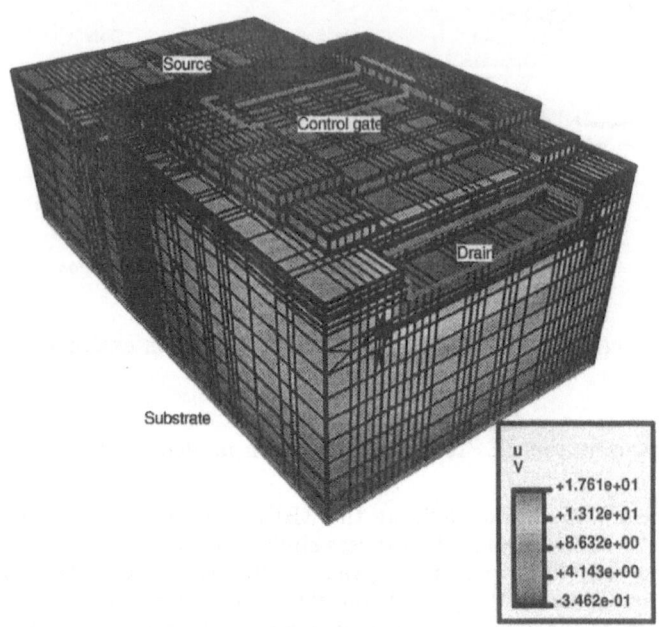

Figure 10: Electrostatic potential during an erase operation

the model optimization resulting from 1D simulations for our 2D process simulations. The only thing we had to take care of was the gridding at the surface of the tunnel implant region in order to describe the segregation and dopant loss exactly. For this region we used an additional refinement step during the adaptive grid refinement.

8. ACKNOWLEDGMENTS

This work was partly supported by the Swiss Foundation for Research in Microtechnology (FSRM), EM Microelectronic-Marin Switzerland, and the ESPRIT project PROMPT (8150) and DESSIS (6075). The authors would like to sepecially acknowledge contributions from Prof. Baccarani and his group at the University of Bologna, Dr. C. Lombardi (SGS-Thompson, Italy), Dr. van Staa (Bosch, Germany), Dr. S. Jones (GMMT, UK), Dr. C. Kalus (Sigma-C, Germany), Dr. B. Baccus (ISEN, France), and J. Lorenz (FhG IIS-B, Germany).

References

[1] C. Hegarty, T. Feudel, N. Hitschfeld, R. Ryter, N. Strecker, M. Westermann, and W. Fichtner, "An Approach to Three-Dimensional VLSI Process Simulation,"

Proc. Process Physics and Modeling in Semiconductor Technology, vol. 93-6, pp. 565-575, 1993.

[2] T. Feudel, "TESIM-4 User's Guide," ISE AG, Zürich, 1994.

[3] N. Strecker, "The 2D-Process Simulator DIOS," ISE AG, Zürich, 1994.

[4] M. Westermann, N. Strecker, P. Regli, and W. Fichtner, "Reliable Solid Modeling for Three-Dimensional Semiconductor Process and Device Simulation," *Proc. NUPAD V (Honolulu),* pp. 49-52, June 5-6, 1994.

[5] G. Garretón, L. Villablanca, N. Strecker, and W. Fichtner, "A new approach for 2-d mesh generation for complex device structures," *Proc. NUPAD V (Honolulu),* June 1994.

[6] N. Hitschfeld, *Grid Generation for Three-Dimensional Non-Rectangular Semiconductor Devices.* Ph.D. Dissertation, Integrated Systems Laboratory, ETH-Zurich, 1993.

[7] S. Müller, *An Object-Oriented Approach to Multidimensional Semiconductor Device Simulation.* Ph.D. Thesis, Integrated Systems Laboratory, ETH-Zurich, 1994.

[8] M. Posselt, "CRYSTAL-TRIM and its Application to Investigations on Channeling Effects During Ion Implantation," *Radiation Effects and Defects in Solids,* vol. 130-131, pp. 87-119, 1994.

[9] M. Westermann, P. Regli, N. Strecker, and W. Fichtner, "Improving Three-Dimensional Semiconductor Modeling through Layout and Process Flow Analysis," *Proc. ESSDERC'94 (Edinburgh),* pp. 351-354, September 1994.

[10] N. Guillemot, G. Pananakakis, and P. Chenevier, "A new analytical model of the bird's beak," *IEEE Trans. on Electron Devices,* vol. 34, pp. 1033–1038, May 1987.

3D Process Simulation Requirements And Tradeoffs From Industrial Perspective

Marius Orlowski
with contributions from
Michael Hartig, Kevin Lucas, Matthew Noell, William J. Taylor, Karl Wimmer, and Tat-Kwan Yu

Advanced Products and Development Laboratory, Motorola Inc.
Austin, Texas 78721, USA

Abstract

This paper discusses the requirements, challenges, tradeoffs, obstacles, economic constraints, and managerial options for three-dimensional process simulation in the context of present and anticipated technology needs.

1. Introduction

The TCAD in semiconductor technology is subdivided in the following areas: device simulation, photolithography simulation, process simulation, topography simulation, and equipment simulation. Some of the areas are clearly delineated from others, for other the borderline is somewhat fuzzy. Device simulation, which is concerned with the computation of the electrical performance of semiconductor devices, is the best established discipline within the semiconductor TCAD world due to the early emphasis and relative simplicity of the governing equations. Although, presently, not the workhorses of device simulation, 3D device simulators are nowadays widely available. This capability is required in the cases of device cross-talk, latchup and soft error phenomena, (SRAM, DRAM, nonvolatile) cell performance assessment, device heating, narrow width and narrow channel phenomena, and others. The domain of process simulation encompasses all processes within the silicon crystal, such as epitaxy, diffusion and gettering, as well as all processes that are related to intimate reactions at its interfaces such as oxidation, nitridation or silicidation. Topography simulation is concerned with the evolution of deposited, etched, sputtered or polished layers that overlay the bulk silicon crystal surface. Equipment simulation is, in a sense, extended topography simulation that includes the physics and chemistry of the entire chamber in which a wafer is being processed. Finally, photolithography is concerned with the calculation of the aerial images and with the resist development profiles, in which it overlaps with topography simulation.

If one reviews these areas more closely with respect to 3D simulation capability, one can make one striking observation: the least advanced area, as far as 3D simulation is concerned, is process simulation, which is of prime interest in this article. In photolithography, it is true, the status of 3D simulation capability is by no means satisfactory, but here photolithography can draw from the significant preparatory work in other (and better established) disciplines such as optics and electro-magnetic applications. The reason for the predicament of 3D process simulation is at least twofold. First, it is a result of historical development. Device simulation had the first priority in the early days of the semiconductor industry, and process simulation was treated merely as an auxiliary science. As M. Law put it [1], "Lack of work produces slower progress." The other areas of TCAD simulations, such as topography simulation, and equipment simulation are pretty recent newcomers to semiconductor TCAD. These disciplines have been more or less inherited from other engineering disciplines with only a few new basic techniques and methods, if at all, to be added. Photolithography simulation had a distinguished career of its own from the inception of semiconductor processing.

Process simulation was initially modeled along the methodology and efforts used to characterize diffusion in metals and other materials which preceded the semiconductor industry by several decades. However, in the intervening years, device dimensions shrank to the scale of a diffusion length and are now approaching the scale of the crystal lattice distance. This, in addition to some peculiar properties of silicon crystal, genuinely challenged process modeling and simulation which saw itself increasingly to be on its own with no older brother to look to for help. Today, the inadequacies of process modeling, including the predicament of verification techniques are considered to be the major stumbling block for reliable device simulation, for which process results constitute the key input ("garbage in - garbage out"). It is noteworthy that, although it is been said that silicon is the best and most widely researched nontrivial element in the periodic table, exactly this knowledge is the weakest link in the silicon device simulation effort.

The second reason is less obvious, but it seems to me that the lagging behind of process simulation with respect to 3D capabilities is due to the relatively simple geometries such as half-planes or cubes encountered in most semiconductor applications. Also the simple additive nature of diffusion effects as opposed to 3D interference effects in the case of photolithography, for example, makes it more amenable to 1D and 2D approximations. To be more precise, the modeling of the diffusion effects might be very complex, but the effects can be extended in a geometrical way to all dimensions. In fact, the simplicity of the 3D geometry of silicon crystal and the additivity of diffusion effects *still* question the necessity of and thus hamper the 3D process simulation capability development.

Of course, if one is to look with a magnifying glass on the issues of process simulation (as on anything else), one will discover a plethora of challenging effects. From them strong reasons and justifications for 3D process simulation capabilities can be derived and postulated. But this reasoning misses one essential point which has to be debated here. Nobody denies that there are valid 3D applications for process simulation, some of which will be reviewed in this article in section 4. The question is, how important are these effects compared with the estimates that can be done based on intelligent 2D process simulation, and how important are the sought-after capabilities compared with capabilities in related fields. Simply, it is a question of priorities. The assessment of the priorities is difficult to accomplish because, on the one hand, the work in each area is so demanding that researchers have to focus on their fields of specialization. They, naturally acquire a vested interest in these fields. On the other hand, there is hardly anybody with sufficient authority, depth of knowledge, and experience who can impartially review all the areas and prioritize the issues. The workshop, such as this one, offers a much needed forum for an thorough review of the 3D process simulation issues.

2. Brief Survey of Arguments in Favor of 3D Process Simulation

In the following I will review and critique some of the frequently encountered arguments pro and con for 3D process simulation capability. There are reasons to argue that the breakthroughs in 3D, if they are going to happen any time soon, will not be due to efforts to be originated from the semiconductor community. Rather they will be from other engineering fields much more experienced in 3D numerics, i.e. from mathematicians, hydrodynamics engineers, plasma scientists, and chip and computer designers. So why not simply wait until all the necessary ingredients become mature, and concentrate in the meantime on some other challenging projects of immediate importance? The counterarguments put forward are of various kinds. One popular counterargument refers to the ostensibly proven shortsightedness of the industry, which craves for new models and new simulators, only when the immediate and urgent need is perceived, oblivious to its own denials in the past that this area might become important. Naturally, the software developers are then caught off guard, because they are unprepared and feel that they have been led astray. Another argument asserts that all the effects in real life are three dimensional in nature and instances 3D examples of how desirable it would be to simulate the entire SRAM cell, for example, wholesale. Still another argument claims that the continuous scaling to ever smaller device sizes necessitates 3D process simulation capability. One of the most convincing arguments in this arena asserts that 3D process simulation capability is badly needed because for narrow devices, 3D diffusion, oxidation, and segregation effects, cannot, at present, be described reliably. I will defer discussion of this argument to section 3. The idea of *virtual fab* is basis for a further argument more to the liking of visionary semiconductor industry managers. This ambitious project envisions that, some day, not into too distant future, all semiconductor processing will be reliably, and even in a predictive way, simulated on the computer. Moreover, the models are supposed be so well calibrated, the tools so easy to use, that an entire process flow can be successfully simulated not only by an expert simulation person, but also by an application engineer without any extensive training in simulation. To make such a vision become reality, a 3D simulation tools would be necessarily an essential and integral part of it. Otherwise, it would be too difficult for an uninstructed application engineer, to assess the 3D effects by educated guesses, intelligent approximations, and clever choice of appropriate 2D or 1D cross sections. I agree with the conclusion put forward in this argument, but not with its premise. Very likely, for the foreseeable future, the technology will always be ahead of simulation, and both modeling and simulation will have a great deal of difficulty catching up with the technology. The pace of innovation will not slacken but rather accelerate as we approach the sub-tenth-micron regime and introduce new techniques, new equipment, new processes and new materials. On the positive side of the same argument one can argue that, in view of the increasingly prohibitive costs of new fabs, approaching presently 2 billion US$, and likely to rise considerably for the 12 inch fab generation, a major effort in developing simulation tools in general, and in 3D capability in particular, is a timely investment and a well spent capital.

Also the point can be made, that simulation tools are useful not only because of their advanced capabilities but because of the expertise and the creativity of the engineer who is using them. Although we are striving to come closer to the point where applications engineers can use simulators in the mode of an 'automatic pilot' (see Karl Wimmer et al's contribution to SISDEP'95 conference [2]) such exercises can be applied only to older well calibrated technologies, and even in such cases, need constant supervision and updates from dedicated simulation experts. However, it is also true that the technological usefulness of a simulation tool grows with its enhanced capabilities and provides better training opportunities for the engineers who use it.

There is, admittedly, a certain or partial truth to all the aforementioned arguments. But as already intimated at several occasions they cannot be taken at their face value. For example, let us consider the argument of ever decreasing device structures; e.g. a 0.1 μm $\times$ 0.1 μm MOSFET. Note that the thermal budget required to manufacture such a device will be very small. Small thermal budget translates into relatively few elementary diffusion jumps. This is equivalent to say, that the diffusion length for the profile evolution is very small. Do we need a 3D extension of a conventional 2D process simulator to simulate the doping distribution for this device? The number of atoms in the channel that determine the device characteristics are a couple of hundreds, and the number of atoms in source/drain close to the channel a couple of thousands or tens of thousands, at most. This means that it should be feasible to describe the implantation and diffusion of the dopants adequately by an Monte Carlo simulator in a 3D domain without any use of differential equations, but based instead on elementary scattering and atomic jump mechanisms. Because of reduced thermal budget the history of diffusion jumps will be short, making the run times affordable. Presently, 3D Monte Carlo ion implantation simulators are available, but no 3D Monte Carlo diffusion simulator. This would seem to be a worthwhile project which would add an orthogonal simulation capability to the existing ones. To my knowledge, work on such simulation capability has not yet been reported.

The other reservations towards a traditional expansion of process simulation to a 3D capability are based on the perception that 3D simulation tools in other areas are of greater importance than in process simulation. Using again the argument of the semiconductor drive to smaller and smaller devices, from the engineering point of view, it is clear that the chip performance is no longer primarily determined by the transistor level behavior but rather by the parasitic effects related to the interconnect architecture. Here, the precise knowledge of deposition, etching, and sputtering rates are of critical importance. These areas have been already covered by the preceding authors in some detail. In section 4, some of these effects will be reviewed from the perspective of an industrial laboratory and their relevance discussed. From an industrial point of view, it is important to prioritize these areas needs with respect to 3D process simulation. In many reviews of simulation needs and requirements, the need for 3D process simulation is found rather at the lower end of the list of priorities. My own rough assessment of the share of 3D applications in the respective fields would be as follows: in i) process simulation 5%, ii) device simulation 10%, iii) photolithography simulation 15%, iv) topography simulation 30%, v) equipment simulation 35%. These are not hard researched and corroborated numbers; they reflect rather my present perception; but they are, I think, indicative of the situation at hand. Some of the areas and capabilities implicitly covered above have been scarcely addressed. For example, the mesoscopic realm of granular materials, particularly in connection with stress and electromigration issues did not produce to this day, a dedicated simulator to address these issues. We are thus entering the sphere of economic tradeoffs and managerial priorities.

3. Economics of 3D Process Simulation

3.1. General Remarks

Not exclusively, but perhaps stronger than elsewhere, in an *industrial* or an *industry related environment* engineering curiosity, pure scientific interest, and a freewheeling desire to explore are restrained by the unyielding economic principle. The question is whether the always limited resources in capital, time, and expertise are being optimally used for the predefined overall goal. In more detailed analysis, the questions that pose themselves in this context are: What are the payoffs in relation to the expected or incurred costs? What is the relative priority of a given project compared with extant related projects? What is the most meaningful sequence of actions and what is the optimal timing for the project to reach a predefined overarching purpose. This issue presupposes that the field of interest is sufficiently well known and has been adequately analyzed and structured.

From these remarks it can be seen, particularly in cases where alternatives do exist, that a project embedded in a complex structure of a multitude of other issues and projects, becomes less of an engineering or scientific issue, and more of a managerial task. Having said this, it cannot be denied that the economic principle and the managerial approach also bear some dangers. They often stifle thinking into a perpetuation of conventional schemes; they might lack the vision for future needs and opportunities; they might misjudge the proper timing; they can curb or even suppress the engineers' creativity and thus hamper the progress; they might be blind to 'paradigm shifts' even if the latter, already in statu nascendi, are clearly recognizable to some; they naturally, as everybody else, cannot foresee future breakthroughs but may act so as to delay these breakthroughs unduly or even thwart them all together. However, the economic principle, beyond the administering of the available but limited resources, and besides its dark sides itemized above, has also some appealing sides. Because of limited resources, the economic principle forces us to confront the reality in an inventive way instead of keeping on dreaming on the fringes of reality. Where and how to strike the balance between unfocused and fortuitous pursuit of all possibilities and the strict selection of specific topics will remain an open question and a perpetual challenge. What one can do, however, is bring all the relevant issues and considerations to the table, debate them collectively, and make a determination in good faith and using the best judgment presently available. This workshop serves excellently this purpose.

3.2. The Transition from 2D to 3D Process Simulation Capability

In purely arithmetic terms, the challenge of the transition from 2D to 3D process simulation capability resembles that of the transition from 1D to 2D; in each case one additional dimension has to be accommodated. It appears therefore to be almost natural to extrapolate the progress linearly in the space of dimensions and advocate unreservedly the attainment of 3D process simulation tools as the next logical step. However, a closer look reveals that the payoffs of the 2D to 3D transition are lower than the payoffs of the 1D to 2D transition, while the corresponding costs are much higher.

Let us consider the costs and required efforts first. To illustrate the following analysis, let us assume a diffusion example for which each dimension has 100 nodes (or unknowns). In both cases of upgrading the simulation capability from 1D to 2D and from 2D to 3D, the 'inherited' number of unknowns (multiplicand) has to be multiplied by the same number (multiplier) of 100. Note, however, that the multiplicands differ dramatically, 10^2 versus 10^4, for 1D and 2D, respectively. Since present computer performance levels and memory resources - impressive as they are - have still difficulty with keeping up with advanced 2D process simulation requirements, it is clear that requirements for realistic 3D applications of similar quality (physics and resolution) are still out of reach. The conclusion is that the transition from 2D to 3D is much more challenging in terms of numerical resources than the transition from 1D to 2D has been. Note also that in 1D, at least as far as simulators based on differential

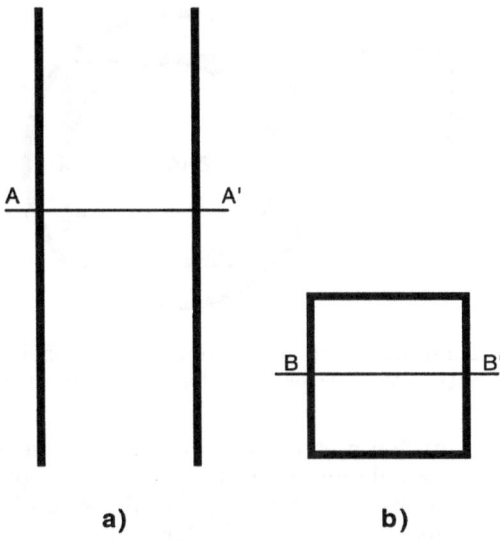

Figure 1: Two implantation regions in top view: a) a strip and b) a square with cross sections indicated by lines $A - A'$ and $B - B'$.

equations are concerned, the meaningful limit of spatial resolution for 1D has already been reached. Today, there is no problem resolving spatial intervals within a lattice constant. Thus, further spatial refinement in 1D is unnecessary and would even be unphysical. This status has not yet been reached in 2D. 3D applications will become, feasible when the grid and solver requirements for 2D cease to pose a serious problem. This status is not likely to be reached in the foreseeable future.

Now, in terms of physics or in terms of engineering insight, I wish to argue that the benefits of the transition from 2D to 3D process simulation capability compared to those from 1D to 2D are meager. The following reasons can be adduced for this assertion. The transition from 1D to 2D was crucial, because in this case, we learned for the first time, how to couple two dimensions with one another. The nature of genuine 2D effects became clear which could not be explained by two orthogonal or multiple 1D simulations. Once this coupling became understood, essentially the same prescription can be exerted by rote to couple an additional, third dimension. In a sense, the latter extension is an exercise of the same concept in a slightly more complex context.

In addition, 1D and 2D simulation capabilities of genuine 3D effects have been hardly exhausted. The nature of many intrinsic 3D effects can be precisely captured by use of 2D cylindrical (r, z), and spherical (r, θ) or (r) coordinates. Yet there is a conspicuous absence of such capabilities in the present mainstream 2D process simulators. It seems natural, to postulate first equipping the present process simulators with such capabilities, before plunging into 3D simulation's troubled waters. 2D simulation capability in cylindrical coordinates for axisymmetric chambers is widely and successfully used in equipment simulation. Since these capabilities for process simulation have been inexcusably neglected, some illustrations are in order at this point.

In Fig.1 two implanted regions are shown; one has the shape of a long strip, and the other of a square. Let us assume that the length of the strip, measured in the units of a diffusion length characteristic for the considered application, is large, whereas the side length of the square is in the same order of magnitude as the diffusion length. If

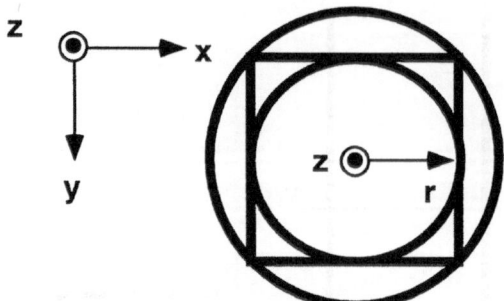

Figure 2: Square-shaped implantation region bounded by an inner an outer circle illustrating 2D approximation of a 3D problem by using cylindrical coordinates.

we were to calculate the diffusion profile along the cross section $A - A'$ and $B - B'$, it is clear that a 2D (x, z) simulation would be perfectly acceptable in the case of the strip, but inadequate in the case of the square; the case of the square poses a 3D problem. But does it require a 3D process simulator to address the problem? Of course, this question can be debated for some time. But the crucial issue is the precision to within which one wishes to predict the diffusion effects. In Fig.2, the square of Fig.1 is bounded by two tangential circles, an inner one and an outer one. Choosing cylindrical coordinates (with z denoting the depth), one can solve precisely the diffusion effects for the two sphere-shaped regions in the cylindrical coordinates (r, z). It is further clear that the solution for the square must be bounded by the solutions for the sphere-shaped regions. Given that our knowledge of the diffusion physics is still rather incomplete and the nature of lateral and transversal profiles little known, it is fair to argue, that the precision of the cylindrical solutions is likely to be good enough for all practical purposes. Obviously, some geometries in real life will be more complex than the highly symmetric shape of a square. But even then, with some inventiveness the problem can be adequately addressed with 2D simulation capability. To illustrate such case consider the elongated rectangle in Fig.3a) representing again the top view of an implanted region. As in the previous example, the cross section $A - A'$ can be handled properly by 2D process simulation in Cartesian coordinates, and the dopant profile across the cross section $B - B'$ near the end of the line needs consideration for 3D effects. In Fig.3b) it is shown how this problem can be divided in subtasks and conquered by 2D simulations separately. The inner rectangle can be characterized by ordinary (x, z) simulations, whereas the squares can be approximated by spheres and handled by 2D simulation in cylindrical coordinates (r, z).

But one can go beyond these geometrical decompositions of a 3D problem and claim that an intelligent pursuit of coupling of orthogonal 2D and 1D simulations has been immolated on the altar of the brute force idol: 3D. To illustrate such an opportunity for coupled 2D simulation modeling of 3D effects consider Fig.4. Although not proven, it seems conceivable that 2D equations in (x, z) and (y, z) planes along the cross sections $A - A', B - B'$, and $C - C'$ coupled in some resourceful way can adequately describe 3D effects. The coupling could be imposed via boundary conditions for outdiffusion problems, or by appropriate diffusivity modeling that would include some kind of geometrical factors, or by adding suitable recombination terms depending on the fluxes in the other dimensions. Exploration of such approaches has been recently initiated [3].

One final argument may further underscore the merits of intelligent 2D approaches as a viable alternative to the heroic efforts to capture the 3D fortress. Let us assume

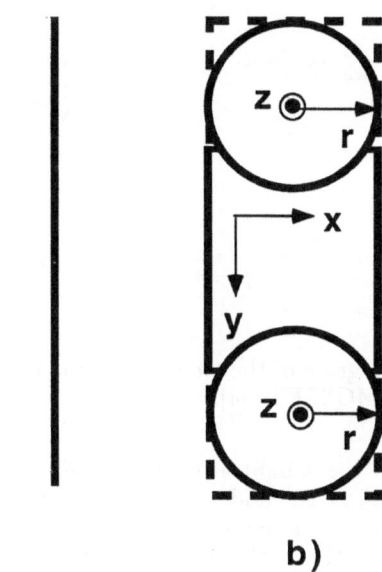

a)   b)

Figure 3: Decomposition of an elongated rectangle into an a smaller rectangle of the same width and two spheres at the rectangle's end. The profile under the inner rectangle can be characterized by 2D simulation in Cartesian (x, y), and the profiles underneath the spheres by 2D simulation in cylindrical (r, z) coordinates.

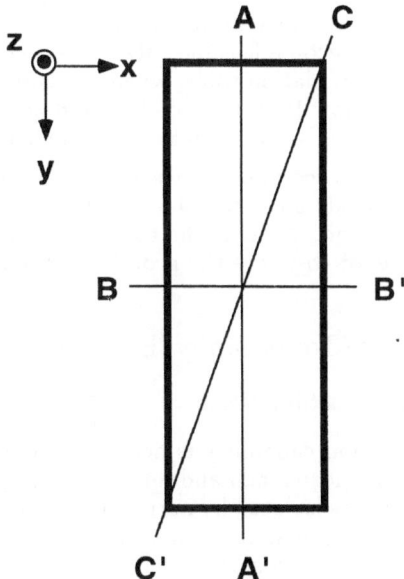

Figure 4: Location of two-dimensional cross-sections for coupled 2D process simulations an approximation of a 3D diffusion problem.

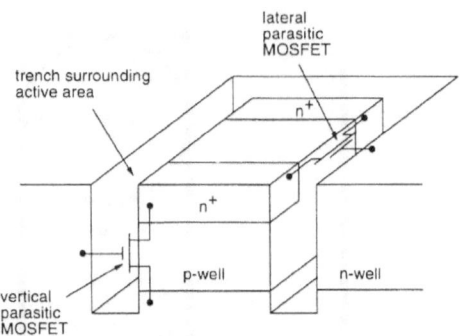

Figure 5: Schematic diagram of the trench isolation technology illustrating the lateral and vertical parasitic MOSFETs [4].

that we have attained the sought-after 3D process simulation capability which can handle $100 \times 100 \times 100 = 10^6$ nodes and several of stiffly coupled, highly nonlinear diffusion-reaction equations reflecting the state-of-the-art in physics knowledge. It goes without saying that this is the most idealistic case. In real life, at least presently and for the foreseeable future, one would have to simplify the physics, reduce the number of equations, remove the most nasty nonlinearities, etc to make a realistic 3D simulation feasible. At the same time, however, the most advanced physics, the same aggressive spatial resolution would be a child's play within the 2D simulation capability (whether in Cartesian, cylindrical, or spherical coordinates). Assuming 2D simulations on a 100×100 grid somewhere from a couple hundred to a couple thousand more meaningful 2D simulations could be computed for *one single 3D* simulation. In *most* process simulation cases the 2D option is presently more attractive than the 3D option. From the above example a simple metric can be derived at which point 3D simulation capability may become feasible. 3D process simulation will enter into the industrial mainstream of simulation tools, when the time for one heavy duty 2D simulation falls significantly below 10 min or so. It would be advisable to concentrate on this task. Any success therein will automatically benefit the 3D numerics.

In summary at the end of this section, we are left with the following provocative questions: Why should we tackle an issue which costs us dramatically more, while paying off dramatically less? and, What is the meaningful metric for the implicated tradeoff?, i.e., At which level of costs are the promised benefits worthwhile?

4. Survey of 3D Process/Topography/Equipment Simulations Issues

4.1. 3D Examples From Process Simulation

Advanced 3D process simulation capability is needed to describe dopant concentrations in edge or corner regions for trench and advanced LOCOS-type isolation technologies. Densely packed devices isolated by narrow isolation structures can mutually influence (transistor cross talk) one another in various ways. In Fig.5 a conventional MOSFET is shown on a silicon mesa surrounded by a trench isolation. Such MOSFET can suffer from parasitic transistor effects indicated also in Fig.5. One frequently encountered parasitic effect, in this context, is the premature inversion of the corner region [4]. In this case, joint action from gate and parasitic gate fields results in highest fields in the corner region and inverts therefore the corner region prior to

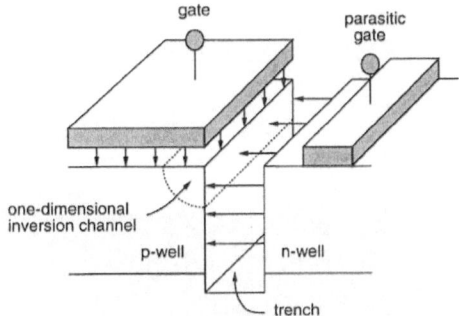

Figure 6: Three-dimensional 'corner' effect. In subthreshold, the electric fields from the conventional and parasitic gate enhance each other, forming a one-dimensional inversion channel on the corner region where the gate and trench oxides meet [4].

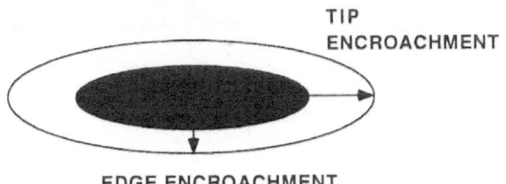

Figure 7: Lateral and transversal encroachment for LOCOS-type narrow oxidation structure.

the inversion of the planar transistor parts, see Fig.6 [4]. Similar effect can occur in absence of parasitic gate but for an underfilled trench. To characterize such devices by means of device simulation a reliable 3D process simulation input is required. The doping levels in the corner region and oxides grown at the corners have to be known precisely. This is a difficult problem; it requires advanced 3D process simulation capability which has to account for 3D oxidation, coupled with 3D segregation (particularly important in the case of boron), and with 3D diffusion effects including OED effects owing to the injection of interstitials from the silicon surface during the oxidation process.

Simulation of isolation structures for ULSI applications seems to be the most urgent and most promising area for 3D process simulation capability. In small LOCOS-type isolation structures the prediction 1) of lateral and transversal encroachment, 2) of the pad oxide punchthrough, and 3) of the stress distribution in the adjacent silicon regions is of critical importance.

1) Encroachment of the oxide body into active silicon region is the most serious drawback of LOCOS-based isolation structures. In Fig.7 an LOCOS-type isolation region is shown in a top view. The filled ellipse indicates the full extent of oxide thickness and the outer ellipse perimeter delineates the extent of the encroachment. It can be seen that the tip encroachment is larger than the edge encroachment. The ratio of this quantities is a complicated function of the aspect ratio (length/width) of the oxidation window as well as of the characteristics of the oxidation process such as temperature, ambient gases, and ramp up and ramp down temperature profiles.

2) Pad oxide punchthrough is already a well-known 2D effect for narrow but still wide isolation structures. This effect is a limiting factor for the use of LOCOS-

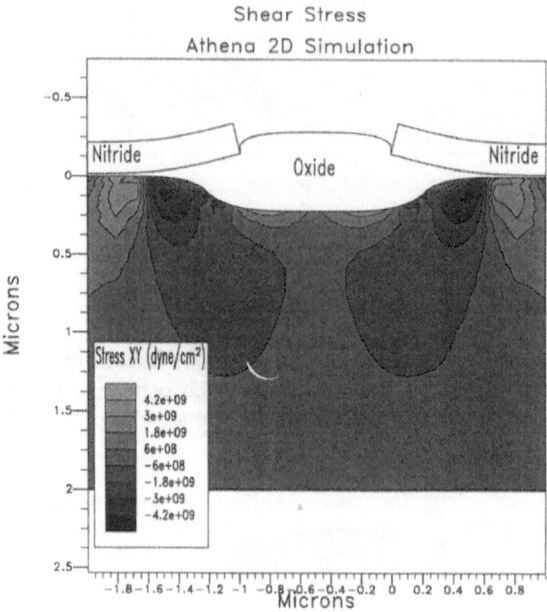

Figure 8: Simulated shear stress distribution for a narrow nitride clad LOCOS oxidation structure [5].

type structures for very aggressive isolation requirements. When the width also is being gradually reduced (in addition to the reduction of the length) the pad oxide punchthrough can occur 'earlier', i.e. at larger lengths of the oxide structures. It is critically important to predict the onset of pad oxide punchthrough effects as a function of isolation structure length for a given width. 3D oxidation capability could greatly aid the layout and acceptable tolerances for the oxide regions. 3D process simulation could provide guidance to optimization efforts in this area.

3) Postoxidation stresses in silicon cause serious leakage problems and other manufacturing uncertainties including gate oxide thinning. In Fig.8 a 2D stress distribution of an nitride clad structure [5] is shown. It can be seen that for a narrow structure the entire silicon area the oxide structure (and not only the regions underneath the oxide bird's beaks) are under stress. However, these 2D stress simulations might be not reliable. In 2D simulations the transversal volume expansion and the associated strains and stresses have been ignored. Only 3D oxidation simulation capability (or 2D simulation in cylindrical coordinates for certain cases discussed in section 3.2 in case of diffusion) could reliably describe small oxide structures. 3D stress distribution is also important for re-oxidized and filled trench structures. Here, in some cases, stresses might reach levels of the yield stress and generate extended crystal defects. Oxidation stresses are particularly detrimental for isolation structures on SOI substrates.

There are other 3D process simulation applications such as silicidation, outdiffusion from narrow and short material layers (islands) and in micromechanic device applications. Silicidation is confronted with similar problems as oxidation for short and narrow devices (silicidation of source/drain regions and of the gate electrode). Outdiffusion is an issue, for example, for emitter technologies when the polysilicon emitter layer is small both in lateral and transversal dimensions. Optimization of micromechanical devices with intrinsically 3D geometries will, certainly, benefit from coupled

Top Down View

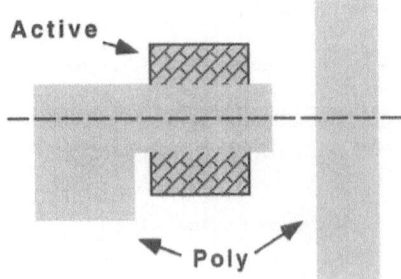

Cut Section on Dotted Line

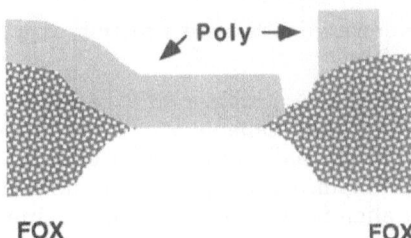

Figure 9: Schematic drawing of optical coupling between adjacent patterned lines over a non-planar substrate caused in this case by LOCOS-type isolation.

advanced 3D process and topography simulation capabilities.

4.2. 3D Examples From Photolithography Simulation

Photolithography is the enabling technology in semiconductor processing which ultimately determines the density of devices on a chip. As the structures are scaled down and the density of devices increases, 3D photolithography simulation becomes a critical capability. Three representative 3D simulation applications will be subsequently discussed to illustrate the kind and scope of 3D photolithography issues. The first application is concerned with patterning of two polysilicon lines and is illustrated in Fig.9. Because of the close vicinity of the two lines, optical coupling between them interferes with the pattern transfer. The optical coupling is aggravated by the underlying nonplanar topography due to the nonplanarity of the LOCOS-type isolation structure. The nonplanar substrate causes also line thinning and notching effects. It is difficult to design proper spacings and width of the mask. A 3D photolithography capability would greatly help with the proper sizing of the mask or with addition of serifs.

The second example is related to the alignment issues. In Fig.10 a top down view and a cross section view of alignment marks is shown. The size of the marks are typically in a couple of micrometers range, which is only few times larger than the alignment wavelength. The alignment marks are cut into the substrate and display therefore non-planar topography. This non-planarity is required in order to get a good alignment signal. The difficulty is to properly interpret the reflected alignment sig-

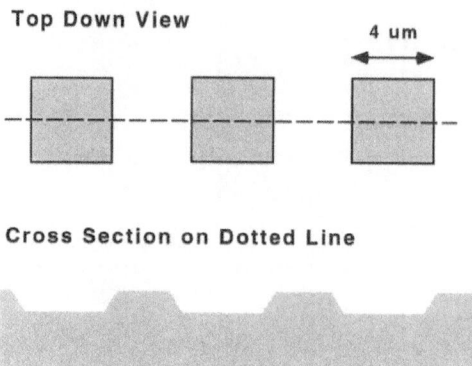

Figure 10: Top down view and cross section of alignment marks.

nal. Therefore, the light scattering problem must be solved rigorously. This requires a 3D photolithography simulation capability. The slope of the mark's topography affects the reflected light signal as does the presence of photoresist or other material overlaying the marks. The main problem, which 3D simulation can help with, is the design or modification of alignment imaging schemes to improve signal intensity and uniformity.

The third application addresses the issues associated with the use of alternating phase shift masks (PSM). The phase shifting effect is achieved by etching of grooves into the glass about one wavelength deep. Such a mask is schematically depicted in Fig.11. Alternating PSMs provide unsurpassed imaging potential (see contribution by K. Lucas to SISDEP'95 conference [6]). However, their usefulness is limited to continuously repeating structures because of the difficulties of designing masks for more general structures. In addition, the scattering of light by the high glass walls of the phase shifting openings causes the light through these openings to be of lower intensity than through the unshifted parts of the mask. The difficulty is the proper design of the termination of the alternating PSM lines. 3D simulation can help to correct the shifted/unshifted intensity bias and to properly size and to place the openings in the regions where three lines diverge.

4.3. 3D Examples From Topography Simulation

A representative example for 3D topography simulation is the etching of vias, covering the opening with a glue layer, and filling with a conductor such as tungsten. The control and uniformity of these steps is crucial for the interconnect reliability issues. In Fig.12 3D current distribution is shown for a four contact via configuration shown in Fig. 13 together with one and two contact configuration. The current values have been extracted from 3D calculations for the the three contact configurations and are shown in Fig. 14. Three observations can be made [7]. First, as expected, the maximum current density is highest for the one contact configuration and lowest for the four contact configuration. Second, the current is highest at the last downstream via contact. And third, the current maximum is reached at the re-entrant corner of the via, a circumstance which can be also seen in the 3D plot of the current distribution shown in Fig.12. Using the current distribution information, one can calculate the

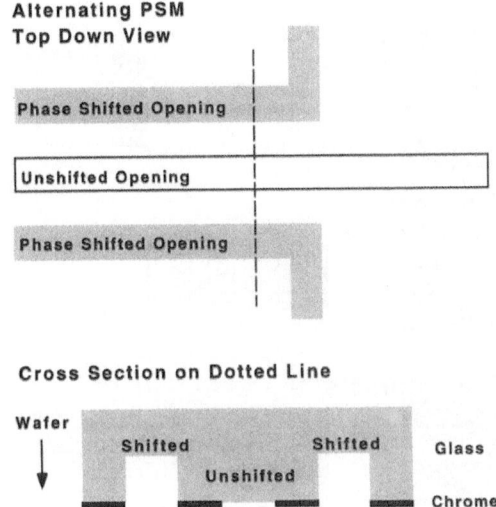

Figure 11: Top view and cross section of an alternating phase shift mask.

CURRENT DENSITY VECTOR

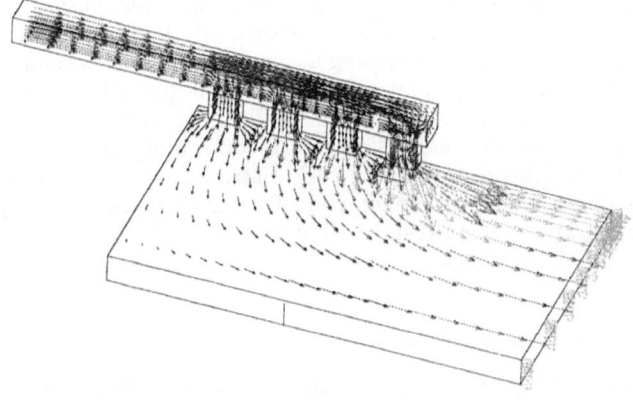

Figure 12: 3D simulation of current crowding effects in a four via configuration to assess the electromigration reliability issues [7].

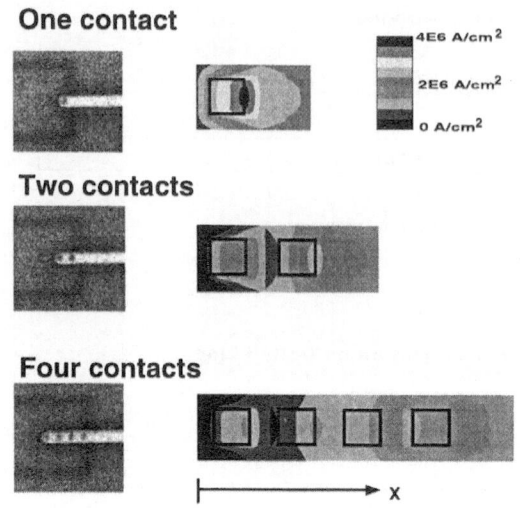

Figure 13: Test structures with 1,2, and 4 W-plug contacts along with current density contour plots from simulations, see Fig.12 [7].

median-time-to failure using Black's equation for the various via configurations. The agreement with experiment in some cases is very good, see Fig. 15.

From the topography point of view it is important to know how sloped is the reentrant corner of the via, and what is the curvature radius of the via plug in various planes. These questions require sophisticated 3D topography simulation tools including deposition, planarization (reflow or CMP), and etching modules.

Another challenging application are the 3D ion milling and redeposition effects. In Fig.16 a result of a 2D ion-milling simulation is shown [8]. The bottom layer of platinum is ion-milled, but as a result of sputtering also redeposited on the sloped sidewall. In real applications the redeposited platinum can provide parasitic conductive path between top and bottom conductive layers. The challenging question is what happens in 3D corner regions. To describe these effects a 3D ion-milling simulation capability is needed along with the following input: i) 3D distribution of re-emitted material as a function of incident angle, ii) sputter yield for different incident angles of the ion beam, iii) 3D topography (slanted sidewall, curvature radii of the opening) of the trench, and iv) 3D shadowing effects.

4.4. 3D Examples From Equipment Simulation

Characisticly, most 3D equipment simulation applications arise from the departures from the axisymmetry. Axisymmetric problems are ubiquitous in equipment simulation and have been treated successfully in 2D cylindrical coordinates. 3D equipment effects affecting the uniformity of deposition or etching stem from breaking of the axisymmetry by non-axisymmetric supply and removal of process gases. This is the case in some of the RTP, low pressure plasma etching, and plasma enhanced deposition equipment. Additional nonaxisymmetric problems arise from nonuniform surface charging in plasma etching. This results in non-uniformly tilted trenches

Simulated Current Crowding

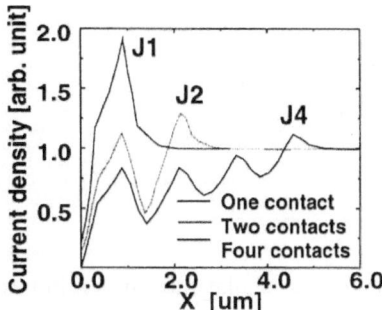

Figure 14: Simulated current crowding along the metal line interface adjacent to the plugs. Current crowding and electromigration issues are most severe at the re-entrant corners [7].

Measured vs. Predicted Lifetimes

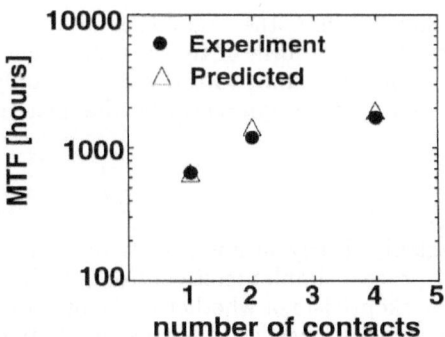

Figure 15: Measured and calculated median-time-to-failure for the three via configurations of Fig.13 [7].

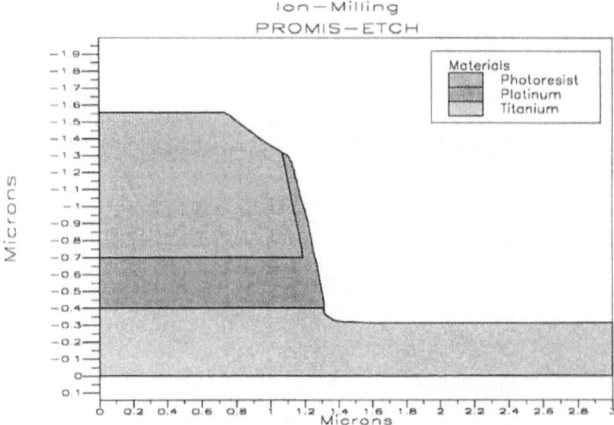

Figure 16: Simulation of ion milling of platinum in conjunction with platinum redeposition on the sloped sidewall [8].

across the wafer. The latter problem requires a coupling of chamber scale, wafer scale, and feature size 3D simulation capabilities.

Departures from axisymmetric geometry also can be caused by buoyancy effects in horizontal furnaces. For horizontal diffusion furnaces, where a gas or a mix of gases is flowed across a wafer boat at temperatures up to 1100 C, gravitational forces may come into play in the gas flow in the form of buoyancy. In this case, a temperature gradient might establish itself along the vertical dimension and alter uncontrollably the chemical kinetics. Because of the thermal gradient across the axis of symmetry, only a 3D simulation of the gas flow may account for these effects. Because of buoyancy effects horizontal furnaces are now being replaced by vertical furnaces.

In the domain of fluid dynamics, 3D recirculation flow patterns create also deviations from axisymmetry.

Finally, for some RTP reactors, while the chamber is of an axisymmetric cylindrical design, the lamps are designed in a form of tubes running across the chamber. The coupling of the radiation from the lamps to the wafer can only be accounted for by 3D models due to the breaking of the symmetry by the heat transfer.

5. Conclusion

Although there are undeniably clearly outlined needs and requirements for 3D process simulation capability, 3D process simulation is confronted with stiff competition from related fields and with the skepticism of whether the value of the anticipated benefits justifies the heroic efforts which are needed to achieve it. As pointed out in this paper, 2D simulation opportunities using cylindrical and spherical coordinates have not been exhausted in present 2D process simulators. At the level of ultra small devices, a more radical shift in simulation capability might be advisable: instead of extending the present 2D process simulators (which are based on differential equations) to a 3D capability, a 3D Monte Carlo simulator of implantation, diffusion and surface reactions could be the more appropriate tool which could also capture the statistical dopant variation. However, there are other compelling reasons to advocate stronger development efforts in 3D process simulation, such as the vision of a *virtual fab*.

Overall, the present level of efforts in this field is presumably adequate when put in the perspective of available resources and of the needs and requirements in other fields.

References

[1] M. Law, Proceedings **SISDEP** 1993, p.1 (1993)

[2] K. Wimmer et al, **SISDEP'95** (1995)

[3] K. Wimmer and M. Orlowski, in preparation

[4] M. Noell et al, **SISDEP'91**, p. 331 (1991)

[5] J. Pfiester et al, **Symp. VLSI Technology**, p. 139 (1993)

[6] K. Lucas et al, **SISDEP'95** (1995)

[7] H. Kawasaki, Ch. Lee, T.-K. Yu, **Thin Solid Films**, 253, p.508 (1994)

[8] E. Strasser, Ph.D. Thesis, Technical University, Vienna (1994)

Author Index

Dietmar Schroeder

Modelling of Interface Carrier Transport for Device Simulation

1994. 69 figs. XI, 221 pages. ISBN 3-211-82539-8
Cloth DM 186,–, öS 1298,–. (Computational Microelectronics)

Wolfgang Joppich, Slobodan Mijalković

Multigrid Methods for Process Simulation

1993. 126 figures. XVII, 309 pages. ISBN 3-211-82404-9
Cloth DM 198,–, öS 1386,–. (Computational Microelectronics)

Narain Arora

MOSFET Models for VLSI Circuit Simulation

Theory and Practice

1993. 270 figures. XXII, 605 pages. ISBN 3-211-82395-6
Cloth DM 298,–, öS 2086,–. (Computational Microelectronics)

Wilfried Hänsch

The Drift Diffusion Equation and Its Applications in MOSFET Modeling

1991. 95 figures. XII, 271 pages. ISBN 3-211-82222-4
Cloth DM 164,–, öS 1148,–. (Computational Microelectronics)

Prices are subject to change without notice

Springer-Verlag Wien NewYork

Sachsenplatz 4–6, P.O.Box 89, A-1201 Wien · 175 Fifth Avenue, New York, NY 10010, USA
Heidelberger Platz 3, D-14197 Berlin · 3-13, Hongo 3-chome, Bunkyo-ku, Tokyo 113, Japan

Franz Fasching, Stefan Halama,
Siegfried Selberherr (eds.)

Technology CAD Systems

1993. 199 figures. VIII, 309 pages.
Cloth DM 128,–, öS 896,–
ISBN 3-211-82505-3

Prices are subject to change without notice

The proceedings of the first "Workshop on Technology CAD Systems" is an authoritative tutorial on CAD software systems for the physical design of semiconductor devices and manufacturing processes. Fourteen invited papers by academic and industrial representatives from USA, Europe, and Japan, as well as from commercial software vendors provide an excellent overview of the work in this area. This book covers the following topics: Coupling and integration of process simulation, device simulation, parameter extraction, and circuit simulation, technology CAD requirements, architectures and strategies of existing CAD systems, implementation and software aspects, practical applications and experiences with technology CAD, and directions of future work in this field.

 Springer-Verlag Wien New York

Sachsenplatz 4–6, P.O.Box 89, A-1201 Wien · 175 Fifth Avenue, New York, NY 10010, USA
Heidelberger Platz 3, D-14197 Berlin · 3-13, Hongo 3-chome, Bunkyo-ku, Tokyo 113, Japan